The Rubbish Book

I would like to offer a massive thank you to Ecosurety. Without them the book would not be in your hands right now.

I have worked with Ecosurety for the past twelve years, latterly as CEO and currently as a non-executive director.

Started in 2003, Ecosurety has grown significantly and has worked with Mars, Nestlé, Co-op and Morrisons, to name a few. Ecosurety helps these companies navigate the complex world of producer responsibility. It also creates and develops innovative sustainability programmes and campaigns, focusing on helping the public learn more about recycling and sustainability.

Ecosurety stands out in its industry for its transparent and ethical approach to business and is the only accredited B Corp in the environmental compliance space.

The experience gathered from working with Ecosurety has shaped a lot of this book and I am extremely grateful for their support over the years.

Thanks team!

The Rubbish Book

A Complete Guide to Recycling

James Piper

unbound

First published in 2022

Unbound
Level 1, Devonshire House, One Mayfair Place, London, W1J 8AJ

www.unbound.com

Text Design by PDQ Digital Media Solutions Ltd.
A CIP record for this book is available from the British Library

ISBN 978-1-80018-086-4 (paperback)
ISBN 978-1-80018-087-1 (ebook)

Printed and bound in Great Britain by Clays Ltd, Elcograf S.p.A.

1 3 5 7 9 8 6 4 2

Thanks to Margaret, Iain and Sam for checking
the book was not complete rubbish...
...and to Ellie for putting up with my rubbish.

Books are not widely recycled. Rubbish, I know! The adhesive that binds the pages would contaminate the paper recycling process, unfortunately. So if you have finished reading this book and become a recycling champion, why not give it away? This page allows you to record your name and date of ownership so future readers can see how much use can be had out of just one book.

Name   **Date received**

_____   _____

_____   _____

_____   _____

_____   _____

_____   _____

_____   _____

_____   _____

_____   _____

_____   _____

_____   _____

_____   _____

_____   _____

_____   _____

_____   _____

_____   _____

_____   _____

_____   _____

_____   _____

Contents

Introduction

Everything can be recycled.

Everything!

In the world of waste, there is a big difference between whether something can physically be recycled, and whether it actually is. It all comes down to the thing that dictates so many of our issues: money.

Chocolate wrappers, crisp packets, polystyrene... they can all be recycled. I've seen it happen. I've watched a cat food pouch become oil ready to be made into new packaging. This surprises people, as you will often hear in the media that 'product X cannot be recycled'. This needs to be adjusted by one word: 'product X cannot be recycled *economically*'.

The economics of waste is fascinating. There are whole teams within companies whose job it is to assess how to send us, their consumers, products in the most economically efficient ways. Ecommerce businesses do not really want to send us small items in large boxes – why waste the box? However, standardised packaging and posting is much more efficient than designing a different box for each product, and the environmental cost is nearly always a secondary consideration to efficiency.

However, the world has become a very different place in the last few years. Now, many companies have realised that efficiency should take a back seat to the environmental story they must tell. It feels different out there. More than ever before brands and retailers *want* to do the right thing.

Unfortunately, consumer emotion often moves faster than a supply chain can evolve, so shortcuts are taken, and positive PR sought, as the world shuns plastic and turns to alternatives without questioning the environmental impact of the new options. These knee-jerk reactions can cause more damage than the problems they seek to solve.

For me, this came to a head on 17 October 2018.

I scroll through BBC News most days with a cursory glance at headlines. On that day, a BBC article appeared on my feed which grabbed my attention. The headline read, 'Plastic pollution found on shipwrecks'. I thought to myself, 'How could this happen? A modern invention on historical shipwrecks? Why is plastic everywhere? We need to get rid of it!'

So, I clicked the link to the article. There was a picture, and sure enough, surrounding the waterlogged cannons of HMS *Invincible* in the Solent sat nine plastic bottles, roughly two litres in size. But wait... surrounding the nine bottles were thirty-eight aluminium cans. THIRTY-EIGHT! And yet the headline did not say 'Packaging found on shipwreck' or 'Mostly metal, some plastic found on shipwreck' – it referred exclusively to plastic. Fed up with this bias, I sat down to write a rubbish book, with the aim of explaining the nuances of the recycling world and debunking the myths. The result is the book currently in your hands.

The Rubbish Book was born out of my first-class ability to frustrate my friends by giving them advice on which bin to put things in and how to dispose of items properly. After teaching one of my friends about putting caps back on bottles before recycling, which then led to a lengthy conversation about other things he had not been getting right about recycling, he said, 'Someone should really write a book on this stuff.' I think it was mainly in an effort to get me to stop talking. But I wrote the book anyway.

I have worked in the recycling space for over a decade and been the CEO of a leading environmental consultancy for the last five years. Over that time, I have learned so much about the world of waste – and I have also learned we do not communicate things very well. Messaging to consumers about packaging and recycling is complicated and disjointed.

Here's an example. In order to share tips and tricks, I have recently joined Instagram, where I was alerted to just how many people who are 'anti-plastic' have switched to buying toilet roll from online companies that do not use plastic packaging. However, these companies individually wrap virgin paper toilet rolls in paper and the largest manufacturers ship their rolls from China. As an alternative, leading supermarkets sell environmental branded toilet roll that is made from paper recycled in the UK but wrapped in a few grams of plastic, which can actually be recycled by taking it to large supermarkets (page 159). The environmental impact of buying rolls individually wrapped in paper, compared to a multipack wrapped in plastic, is huge, and that's without even considering the journey from China that each package must undertake.

You see, the problem with plastic is actually a problem with packaging. And the problem with packaging is actually a problem with consumerism more broadly. For example, our desire to buy fresh products at any time of the year, with no consideration of seasonality, means fruit and vegetables have to be shipped around the world – a journey they will only survive with packaging. So in order to *really* effect environmental change, we must question the way we buy and consume, rather than what something is wrapped in.

Consider the money in your pocket. In 2016, the UK began rolling out banknotes made of plastic (polypropylene, to be precise), yet I have never been fortunate enough to find a plastic banknote in the street, on the side of a road, or littering our beaches – a stark contrast to widely discarded packaging. Why not? Money has a perceived value that's much higher than our packaging, yet its real material value is the same. It is interesting then that we hold packaging in such low regard. If we valued it the same as the money in our pocket, perhaps we would see less litter and more care.

I get it: plastic feels wrong. It is hard to change that perception, and in lots of ways it is indeed wrong. The bad management of waste plastic leading to plastic in the ocean, the idea that we can export away our waste problems... wrong! But I would add that most packaging has issues, whatever it is made of. Why does a toothpaste tube come in a cardboard box? Why is a paper straw acceptable when it is not needed? We should challenge our societal aims and focus on what is important, reducing *all* packaging and reusing wherever possible.

Having said all this, change takes time, and both packaging and recycling have an important place in society. Therefore, it is essential we understand what can be recycled economically and what cannot, and where possible ensure we avoid packaging that cannot be collected or recycled. We must also demand of our governments that low-quality materials, which have little economic value in the recycling chain, should not be sent for recycling abroad, where they are rarely recycled or well managed.

This is the purpose of this book: to teach you all the tips and tricks I have learned from years spent working with brands and retailers helping them improve what they sell you. I believe that if we are empowered with recycling knowledge, together we can demand change and ensure as wide a range of materials as possible are collected and recycled. With confidence, you can do your bit to help make the world a better place.

This book will not tell you how to cut plastic out of your life. It will not tell you how to become self-sufficient. There are plenty of really interesting books for that. This is a practical guide for people who want to tweak their lifestyle by using facts and knowledge to improve their environmental footprint.

Thank you for buying this book and supporting this goal. I hope that you will find it a useful and practical guide to the world of recycling.

Good luck on your sustainability journey.

How to use this book

The Rubbish Book is designed to be a complete guide to recycling, with answers to 150 questions about the world of waste. The book can be read in one go or simply referred to when you need to know the answer to a specific question.

One key thing to remember is that unfortunately all local authorities in the UK recycle differently, so you will need to check how your particular area collects and recycles – your local authority website is the best place to look for this information. Always remember, the guidance in this book is current good practice, not advice on your local area.

The order of the contents is designed to build your knowledge as you work your way through the book. The sections are:

- **The Rubbish Basics** – terms and definitions that are useful later in the book.
- **The Rubbish History** – how the packaging and products discussed most in the book were developed.
- **The Rubbish Collection, Sorting** and **Recycling** – the next three sections cover the whole process of recycling, focusing on each material and product.
- **The Rubbish Knowledge** – answers to frequently asked (and some rarely thought about!) questions, designed to take the mystery out of the recycling process and arm you with some useful tips and tricks.

- **The Rubbish Symbols** – key symbols you will find on packaging and products explained.
- **The Rubbish Encyclopedia** – the recyclability of fifty household items in alphabetical order, with explanations of why they can or cannot be recycled.
- **The Rubbish Problems** – the things we read about in the news: plastic in the ocean, banned packaging, microplastics – this section looks at it all in detail.
- **The ~~Rubbish~~ Better Future** – it's not all bad! There are some amazing initiatives and changes coming in the recycling space; this section explores some of them.
- **Glossary** – a full list of terms and jargon defined.

It is in equal measure fascinating and frustrating how quickly recycling data and information changes. For example, as I wrote the book, plastic waste being exported to other countries dropped by 35 per cent, so as quickly as I could write a section on exported waste it would be out of date. For this reason the book has been carefully written to provide you with the best overview, but avoid details that would be out of date by the time you read it.

However, this information is important so there is a companion website to the book which will be populated with data and any updates – please visit therubbishwebsite.com.

1

The Rubbish Basics

Before we get going on our recycling journey, you are going to need some of the basics; it will help make sure we are all using the same language throughout the book. Legislation, definitions, meetings, this section has it all!

What is recycling?

Creating a 'complete guide to recycling' means starting with the obvious question.

Too simple? Maybe not…

What a waste!

According to the Oxford English Dictionary, the formal definition of recycling is 'the action or process of converting waste into a reusable material'.[1] This simple definition is more complicated than it first appears. For example, if you shred an old envelope at home and then put the shredded paper into a box as packing material, does that make you a recycler? Is cleaning an old glass jar and using it to store coins a recycling activity?

Ultimately, this comes down to the definition of 'waste' and when something has become a new 'reusable material'. This is important in the recycling industry as the definition of waste and when something is recycled affects whether the item falls within waste legislation and is counted towards recycling targets.

According to the EU Waste Framework Directive, 'waste' is defined as 'any substance or object which the holder discards or intends or is required to discard'.[2] Legal guidance suggests waste is likely to be left over and unwanted or can no longer be used for the original purpose.

An item will be considered 'reusable material' if it meets an 'end of waste' test and is no longer considered waste. This means the item has become a completely new product, which has a market, and is different from the original.

Based on these definitions, shredding an old envelope to fill a cardboard box does constitute recycling; the envelope would have been discarded and has been converted (shredded) and put to a new use. The glass jar has not been recycled, as it is being used for the same purpose (to store something) and is not sufficiently changed from its original use (a glass jar).

What is packaging?

A lot of this book focuses on packaging. Other areas are covered including electricals, batteries, tyres and textiles. However, packaging is by far the most collected and recycled waste.

The definition of packaging should be simple. However, like recycling, packaging is also a bit complicated.

All packaged up

According to UK packaging legislation, which has been designed to reduce packaging and increase recycling (page 14), packaging is defined as 'any material used to hold, protect, handle, deliver and present goods'.[3] This includes packaging around raw materials, such as the packaging around ingredients a restaurant might use, right through to finished goods that are sold, for example the takeaway box the food is contained in.

Packaging can include boxes, bags, address labels, pallets, tape for wrapping, rolls, tubes and clothes hangers. This can add a bit of complexity when packaging is sold to a consumer as a product. For example, wrapping paper which has been wrapped around a gift at the checkout is packaging, but a roll of wrapping paper sold as a product is not considered packaging.

In the packaging industry, companies refer to primary, secondary and tertiary packaging. Primary packaging is found directly around a product, such as a wrapper around a chocolate bar. Secondary packaging contains a group of products, such as a box which contains multiple chocolate bars to be displayed on a supermarket shelf. Tertiary packaging is the transit packaging used to move products around, such as a pallet which carries multiple boxes of chocolate bars to the supermarket.

What makes up your packaging?

Packaging as a category is made up of specific materials. In the recycling industry, they are usually listed in the same order: paper or cardboard, glass, aluminium, steel, plastic and wood.

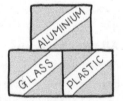

Paper and cardboard

Widely used due to its low cost and ability to hold its shape, paper and cardboard are ideal for transporting products.

Corrugated cardboard is becoming more frequently used due to a relatively recent increase in home deliveries as a result of a shift to online shopping.

Glass

Glass keeps moisture and gas out, so it is ideal for food contact applications, such as jars and bottles. Another reason for its popularity is that its transparent nature allows consumers to see products before purchasing them.

Aluminium

Aluminium is most frequently used in the manufacture of drink cans, crisp packets and chocolate wrappers. Aluminium cans are generally lighter than their steel equivalent (around one-third of the weight) and they will not rust or corrode. You will find a thin layer of aluminium in some packaging, such as pet food pouches.

Steel

Steel is mostly used in tin-coated steel cans (which is why they are called 'tin cans'). Tin cans are generally used to contain food items such as soup and baked beans.

Plastic

There are many types of plastic, from bottles to flexible film. Plastic is the most versatile material, but this versatility brings complex and inconsistent recycling, which is why plastic is often criticised as a packaging material.

Wood

As a packaging material, wood is mostly used for pallets and crates, which move products between warehouse and stores before they get to consumers.

Where do the materials come from?

Paper, cardboard, glass, aluminium, steel, plastic and wood are the main materials used in packaging. But where do these materials come from, and what does recycling them achieve?

Paper and cardboard

Paper and cardboard usually start life as cellulose, the part of the trunk inside a tree that gives it strength. The trees are cut down and the cellulose fibres pulverised and mixed with water. The resulting 'pulp' is pressed and dried, ready to be rolled into paper or cardboard.

This process is almost identical to the recycling process (page 84) but as it involves cutting down trees, it contributes to deforestation. Recycling paper and cardboard reduces the need to cut down trees, but it does shorten the tiny fibres that make up paper, reducing the quality each time it is recycled.

Glass

Glass is made from sand – the same sand you would find at the beach (which is mainly silicon dioxide). Two other ingredients are added: soda ash (sodium carbonate) and limestone (calcium carbonate). When glass is made, collected waste glass can be added; the resulting glass is defined as having recycled content in it. Very usefully, and perhaps something that explains the high recycling rate of glass, adding recycled glass to the process lowers the melting point required to make glass.

To make glass, the ingredients need to be melted into a liquid, which only happens at a high heat. The liquid sand is then cooled and becomes clear as it cools rapidly. Interestingly, glass is never a solid, it is amorphous, which means it sits somewhere between liquid and solid. The molecules that make up glass are not arranged in a regular pattern, as they are in metals. Does this mean your windows could technically 'melt' downwards with gravity? Yes, over an extraordinary length of time they could – but it would take longer than the predicted existence of the universe, so no need to worry just yet.

Aluminium

Aluminium is the most abundant metal in the Earth's crust, making up approximately 8 per cent of its overall mass![4] Aluminium is extremely reactive, so it is very rarely found in its pure form. It is found in bauxite, a sedimentary rock that contains high levels of aluminium.

Bauxite is processed to produce aluminium oxide, and then electrolysis is used to extract the aluminium. The mining and extraction process uses a significant amount of energy; it takes 4 to 5 tonnes of bauxite to produce around 2 tonnes of aluminium oxide, which in turn will produce around 1 tonne of aluminium. Aluminium recycling is a much more efficient and cost-effective way of obtaining this precious metal. Recycling one aluminium can saves 95 per cent[5] of the energy that would be required to extract the metal from raw materials.

Steel

Steel is an alloy of iron and carbon. Iron ore is mined and then smelted in furnaces which blow oxygen through the molten iron, reducing the carbon content of the alloy. Chemical agents are added throughout the process to ensure the removal of impurities. Improvements and advances in the steel process have significantly reduced the number of people required in its manufacture. Between 1920 and 2000, labour requirements have fallen by a factor of 1,000.

Steel cans containing food are plated with corrosion-resistant tin. Therefore, steel cans are known as 'tin cans'. Ordinary metals would rust over time or corrode from the acids that the food inside contains, releasing molecules that could contaminate the food and weaken the can.

In bad news for people switching from plastic bottles to cans to go plastic-free, aluminium cans and steel tins will usually have a thin layer of plastic lining, to prevent their contents reacting with the metals. However, this plastic is so thin and light that it burns off in the recycling process, so it does not inhibit the recyclability of cans.

Plastic

Plastics are polymers, chains of repeating molecules linked together. Due to this structure, most plastic names start with the word 'poly'. As plastics are formed from molecular structuring, there are natural plastics, like tar, rubber or latex. Thermoplastics can be shaped when soft through heating and then hardened to retain a shape when cool.

Plastic manufacturing involves chemically joining single molecules, called monomers, together, which is usually started by distilling crude oil in a refinery. Crude oil will separate into components called fractions, which are different-sized chains of hydrogen and carbon (hydrocarbons). One of these fractions is 'naphtha', which is a building block of plastic. Naphtha is 'cracked' into ethylene or propylene. Using heat and a catalyst, ethylene joins to form polyethylene, and propylene joins to form polypropylene. Most plastics we use are derived from these two polymers.

Plastic is a by-product of the oil industry, which explains its relatively low recycling rates: often, it is cheaper to buy new plastic than it is to recycle it. The economics of plastic recycling have always been its biggest challenge. Future legislation will help increase plastic collection and recycling as it will introduce additional costs for companies that do not use recycled plastic in their packaging (page 216).

Wood

This one is obvious, sorry... wood comes from trees. There are two types of wood: softwood and hardwood. Softwood, such as pine, is made from trees that keep their leaves all year round, otherwise known as evergreen. Hardwood, such as oak, comes from trees that lose their leaves, termed deciduous.

Pallets, the most common kind of wood packaging, can be made of a mix of the two types of wood, with oak or another hardwood for the parts of the pallets that will take weight and pine for the non-load-bearing parts. The wood that a pallet is made from will depend on the type of wood available in its country of origin.

How does rubbish make money?

When you throw rubbish into the recycling bin, it is easy to forget that every company that ends up sorting and processing it later will make money from it.

Who pays for it, and why is your waste valuable?

Another person's treasure

In general, every step in the recycling process adds value. Each company along the chain pays more for the material than the last. In a simple and predictable world, this would be straightforward; however, complication is added as the 'recyclate' is a commodity with fluctuating value.

For example, let us imagine the recycling chain involved with a tonne of plastic bottles, a relatively simple and valuable material. As consumers, we separate and clean our plastics and place them in a recycling bin. The plastic is collected at the kerbside as a mixed stream of plastic types; in the future, plastic bottles are likely to be returned via deposit returns (page 56).

The plastic is transported to a sorter, which uses technology to sort the different plastics by their chemical make-up and colour (page 71). This sorter has bought a mixed load of plastics from the local authority and separated out a tonne of clear plastic bottles. This sorted stream of plastic bottles is worth significantly more than the mixed load.

The bottles are sent to a recycler that shreds and melts them into pellets. A manufacturing company will buy the pellets to put into new bottles, which will create a certain percentage of recycled content in the bottle. As more brands pledge to add a higher percentage of recycled plastic into their bottles, this drives up the value of the recycled pellets. This will push up the price along the chain in turn, as the mixed load and the sorted bottles are in higher demand.

Which is the 'best' material?

This is the hardest and most complicated question in the book. It is easily the question most asked by the public (and quite a few packaging professionals). If there was a single right answer, you can guarantee all companies would be using that material.

The difficulty in answering this question stems from the fact that there is no clearly defined version of the word 'best'. Do we mean cheapest? Most environmentally friendly? Most protective packaging? The look and feel of the item? It's complicated…

It's not easy being green

Recycling is the process of turning a waste product back into a new usable material or product. The value of this recycled product is then intrinsically linked to the worth of the original raw material. As an example, the price of a recycled plastic pellet which will be used to make a PET bottle (page 122) will be measured against a plastic bottle made from the equivalent virgin material. It is this value that determines the likelihood of something being recycled.

Consider the mobile phone in your pocket. At the end of its life, what do you do with it? The vast majority of people will sell it to someone else or return it to a mobile phone recycler or reuse company. The metals in the phone, the electronics, the battery all have a value and are worth something to the person buying it. A leading mobile phone buyer in the UK reportedly receives an average of 150,000 mobiles a month, a return rate created by the desirability and value of the product.[6]

If we think about that in terms of packaging materials, the ones that have a high value, such as aluminium and steel, are more likely to be recycled. This is partly because the equivalent virgin materials are also expensive. However, they are more expensive because of the energy and processes required to produce them. Glass has a high recycling rate in part due to the fact that using it in the recycling process will lower the temperature needed to melt glass. This saves the manufacturer money and so drives the behaviour to collect the material.

Paper and cardboard have a high recycling rate due to how common they are in the home – think of cardboard boxes or newspapers. They are rarely used in on-the-go packaging, as plastic is. Collection is important to the recycling rate of a material, and with a significant amount of plastic being used while we are out and about, it becomes very difficult to collect and recycle.

Economics drive so much of the collection and therefore the recycling rate. Paper and cardboard are easy to collect from the home and businesses; recycled glass makes manufacturing cheaper so is more likely to be used; the metals have high material value.

The problem plastic has is its use. Plastic bottles are actually widely collected and recycled and will improve further on the introduction of a deposit return scheme (page 56). However, plastic is used in lots of on-the-go packaging – salad boxes, fruit tubs, yoghurt pots, etc. – and this makes its collection and recycling significantly harder. It also has an environmental legacy; the properties that make it so good at containing food and drink also make it likely to stick around. Remember, though, that anything that replaces it will need to have the same properties, and will therefore be likely to have the same environmental impact.

Unfortunately, there is no perfect material. Materials that require huge amounts of energy to mine or create, like glass and metal, are likely to be collected and recycled to offset this cost but have the largest environmental impact. Plastic, which is abundant in our packaging system, is used in applications that vary massively in end market value, and therefore different uses have differing collection and recycling rates – bottles are high, for example, and plastic film is very low.

Ultimately, what the system requires is investment from brands and retailers not linked to the value of the raw material, and strong government policy. This is where the biggest change is coming. Companies have made commitments to put recycled content in their packaging, which means they have decoupled the value of the recyclate (the material being recycled) and the raw material. The government has announced wide-ranging policy changes (page 216). Combined, these will create much-needed collection and recycling infrastructure for the materials that currently have the lowest value.

What are the laws relating to waste?

The UK, like many other countries worldwide, manages its waste strategy through legislation and recycling targets. These targets vary for each country, and this book will focus on some of the key UK legislation that exists today, although details of future changes can be found on page 216.

As previously mentioned, waste is defined by the EU Waste Framework Directive as 'any substance or object which the holder discards or intends or is required to discard'.

Environmental legislation means companies that handle waste have a legal responsibility to ensure that waste is stored safely on site, collected by a registered waste carrier (page 47), covered by a valid 'waste transfer note', correctly described on transfer documentation and disposed of according to the waste hierarchy (page 96) at a licensed facility.

Duty of care

Duty of care requires the safe management of waste to protect human health and the environment. The requirements apply to companies that import, produce, carry, keep, treat, dispose of or have control of waste in the UK.

Moving waste

To comply with regulations on moving waste between companies, the company transporting the waste must fill in a 'waste transfer note' which details where the waste has come from and where it is going to. The company must also be registered as a 'waste carrier' (page 47). Interestingly, as this applies to businesses, if you ran a business from your home you would need to ensure the waste related to your business was collected correctly by a company with a waste carrier licence that produced a transfer note.

If the waste collected is hazardous (such as batteries, chemicals or pesticides) the load must also have a 'hazardous waste consignment note' which details the specific requirements for the waste stream.

Waste permits

The Environment Agency issues permits for companies that wish to use, recycle, treat, store or dispose of waste. These permits will dictate the type and volume of waste which may be handled.

Packaging essential requirements

This is the piece of legislation that prevents over-packaging and includes safety requirements around packaging use; however, it is not widely known about or understood by companies. Basically, it requires all companies to use packaging appropriate to the product. Unfortunately, this is down to consumer acceptance in the UK, whereas some other countries dictate how much air space is allowed around a packaged product. In the UK, this legislation is regulated by local trading standards departments, which means you can let them know if you receive something that you consider has too much packaging, and they will then investigate. Most companies will adjust their packaging to avoid prosecution if a complaint is made.

End of Life Vehicle directive

Vehicles sit in their own category. There exists a piece of legislation to assist with take-back of a car at the end of its life; this requires all car manufacturers to have a take-back route for their cars when they are old. As a consumer, you can take your car to a scrapyard and it has to be taken in and disposed of free of charge. Technically, this regulation is a form of 'producer responsibility' (i.e. the producer of the product pays), as described below, although it is rarely classed officially as such.

Producer responsibility

For packaging, waste electricals and batteries there are specific sets of legislation which have introduced the concept of 'producer responsibility'. These regulations ensure that some of the cost associated with collection, treatment and disposal are covered by the manufacturer, brand or retailer of the product.

This is the area that will be most affected in the future with the introduction of 'extended producer responsibility' (page 216).

For now, the three main regulations are:

Packaging

The oldest and most complicated of the three, packaging producer responsibility was enshrined in law in 1997. This legislation requires any company with a turnover above £2 million and that uses more than 50 tonnes of packaging a year to register. This means a whole supply chain (manufacturer, brand owner and retailer) will register and share the cost of recycling packaging. Each year recycling targets are set by the government, and these companies help finance the recycling of packaging to hit the targets.

Waste electricals

Waste electricals producer responsibility became law in 2007. If a company imports, manufactures or rebrands electricals, it must register as a producer and ensure the recycling target for the tonnage it places on the market is achieved. Waste electricals legislation is different from batteries in that the recycling rate is broken down by category of electricals (page 73). If the recycling target for waste electricals is not met by the company, it must pay a compliance fee set by the government.

Batteries

Producer responsibility relating to batteries was introduced in 2009. If a company imports or manufactures batteries, it must register as a producer and pay for 45 per cent of the tonnage it places on the market to be recycled. The battery collection bins you see in shops and supermarkets are funded by this legislation (page 61). Each year millions of batteries are collected and recycled, funded by the companies that put them on the market in the first place.

What are the Sustainable Development Goals?

In September 2015, the UN convened to adopt the 17 Sustainable Development Goals, with a vision of 'leaving no one behind'. These goals, with targets to be achieved by 2030, are intended to transform the future development of the world. Each goal has between eight and twelve targets, with quantitative indicators to measure progress.

The Sustainable Development Goals are a framework for countries and organisations and act as a 'blueprint to achieve a better and more sustainable future for all'.[7] However, they are not legally binding and there are no penalties to countries for failing to meet them. In 2019, the UK produced its first voluntary national review, a 235-page document summarising progress to date.

The 17 Sustainable Development Goals[8] are:

1. No poverty
2. Zero hunger
3. Good health and well-being
4. Quality education
5. Gender equality
6. Clean water and sanitation
7. Affordable and clean energy
8. Decent work and economic growth
9. Industry, innovation and infrastructure
10. Reduced inequalities
11. Sustainable cities and communities
12. Responsible consumption and production
13. Climate action
14. Life below water
15. Life on land
16. Peace, justice and strong institutions
17. Partnerships for the goal

While the goals largely focus on economic and equality targets, there are several targets relating specifically to the environment and waste management. Out of the 169 targets, 17 per cent relate directly to environmental issues (the same as the total number of goals, easy to remember!).

Over the next couple of pages, you will find a summary of the goals that have environmental targets.

Goal 3 – Good health and well-being

Substantially reduce the number of deaths and illnesses from hazardous chemicals and air, water and soil pollution and contamination.[9]

Goal 4 – Quality education

Ensure that all learners acquire the knowledge and skills needed to promote sustainable development, including education for sustainable development and lifestyles.[10]

Goal 6 – Clean water and sanitation

Improve water quality by reducing pollution, eliminating dumping and minimising the release of hazardous chemicals, halving the proportion of untreated wastewater and substantially increasing recycling and safe reuse globally.[11]

Goal 7 – Affordable and clean energy

Increase the share of renewable energy and facilitate access to clean energy research, focusing on renewable energy, efficiency and cleaner fossil fuel technology.[12]

Goal 8 – Decent work and economic growth

Improve global resource efficiency and aim to separate economic growth from environmental damage, with developed countries taking the lead.[13]

Goal 9 – Industry, innovation and infrastructure

Upgrade infrastructure and retrofit industries to make them sustainable, with increased resource-use efficiency and greater adoption of clean and environmentally sound technology.[14]

Goal 11 – Sustainable cities and communities

Reduce the negative environmental impact of cities on the population, giving attention to air quality and household waste management, including improving recycling and reuse systems.[15]

Goal 12 – Responsible consumption and production

Ensure good use of resources, improve energy efficiency, sustainable infrastructure and access to basic services, green jobs and ensuring a better quality of life for all.[16]

Goal 13 – Climate action

Improve education and raise awareness on climate-related issues and ensure climate change measures are built into national policies.[17]

Goal 14 – Life below water

Reduce pollutants in the ocean, particularly those generated from land-based activities, and sustainably protect and manage ocean ecosystems.[18]

What are the COP meetings?

COP stands for 'conference of parties' and refers to a meeting held with all signatories of the United Nations Framework Convention on Climate Change (UNFCCC). The meetings have been held every year since 1995,

although the 2020 meeting, which was meant to be hosted by the UK, was delayed by a year due to Covid-19. The purpose is for countries to report on progress towards climate change targets and for countries to unify around ambitions.

A fair COP

The first COP meeting was held in Berlin, Germany in 1995. Many frameworks and targets have been agreed from the COP meetings, perhaps the most notable being the Paris Agreement (page 23) in 2015, when 195 countries attended COP21.

COP3 took place in Kyoto, Japan in December 1997 and saw the adoption of the Kyoto Protocol, which commits countries to reduce their greenhouse emissions depending on their economic development.

Just in case you ever find yourself in a sustainability pub quiz (is there such a thing?), the most boring bit of trivia that could come up is that COP6 is the only meeting to have been held twice. It was originally held in The Hague, Netherlands in November 2000. However, agreements were not reached on how to support developing countries and enforce targets, so a second meeting was scheduled in 2001, to be held in Bonn, Germany, a few months before COP7.

The COP meetings are becoming more important than ever, as the climate crisis requires countries to work together more than ever before. The 2019 meeting (COP25) held in Madrid actually lasted for two days longer than originally planned and was notably attended by the young climate change activist Greta Thunberg, who spoke at the event.

What is the Paris Agreement?

The Paris Agreement was formed in 2015 during COP21. It is a remarkable achievement that 195 countries agreed to climate change targets, which establish limits on average global temperature warming and, in turn, aim to reduce the risks and impacts of climate change.

Réduire le changement climatique

Adopted on 12 December 2015, the Paris Agreement includes commitments from major countries to cut pollution and to strengthen targets over time. It also provides support to developing countries with climate mitigation. This is underpinned by transparent monitoring and reporting. The agreement would not come into force until countries representing 55 per cent[19] of global emissions formally joined, which happened on 5 October 2016.

The Paris Agreement requires countries to reduce greenhouse gas production by increasing renewable energy; keep the global temperature increase well below 2°C, trying to limit it to 1.5°C compared to pre-industrial levels; review progress every five years; and spend $100 billion a year providing climate finance to support poorer countries.

The Paris Agreement replaces the first environmental treaty, the Kyoto Protocol, which was agreed during COP3 in 1997. The Kyoto Protocol was focused on developed countries, so the wider remit of the Paris Agreement should broaden the focus to all countries.

The US is the second-largest global emitter of greenhouse gases. Under Donald Trump's 2017–21 presidency, the US pulled out of the Paris Agreement, citing unfairness in the structure of the agreement. However, following Joe Biden's success in the 2020 presidential election, the US has now re-entered this important agreement.

What is the UK Plastics Pact?

The UK Plastics Pact is a set of targets agreed by most of the UK's largest brands and retailers, aiming to create a circular economy (page 116) and reduce  unnecessary plastics. It is a programme delivered by WRAP (Waste and Resources Action Programme), a UK charity focused on circular economies and education.

A pact agenda

The UK Plastics Pact launched in April 2018, based on an initiative developed by the Ellen MacArthur Foundation, which has set up equivalent pacts in many countries around the world.

All targets in the UK Plastics Pact are supposed to be achieved by 2025, with four main objectives: 100 per cent of all plastic introduced to the market to be reusable, recyclable or compostable; 70 per cent of plastic to be effectively recycled or composted; elimination of problematic single-use plastics; and an average of 30 per cent recycled content in packaging.[20] Signatories to the pact, which include supermarkets, food producers and packaging developers, regularly report on progress towards these targets.

By contrast, the 2018 baseline data showed that the then fifty-five members of the pact sold 1.1 billion items of problematic or unnecessary plastic; only 65 per cent of plastic was recyclable; the national plastic packaging recycling rate was 44 per cent; and the average amount of recycled content in the packaging was 10 per cent.[21]

The objective most open to interpretation is the elimination of problematic single-use plastics. Helpfully, the Plastics Pact set out an initial eight problem plastics.[22] These are: disposable plastic cutlery, all household polystyrene packaging, cotton buds with plastic stems, plastic stirrers, plastic straws, oxo-degradable plastics (page 224), PVC packaging and disposable plastic plates and bowls. Some of these are also covered by the single-use plastic bans (page 207).

This could be just the beginning, with a further nineteen problematic plastics that have been identified for potential reduction and eventual elimination. One thing is for certain: a united voice and approach allows for faster and greater change.

What is single-use plastic?

'Single use' was Collins Dictionary's word of the year in 2018, and it is easy to see why: everyone was (and is) talking about single-use plastic. Usage of the term has seen a four-fold increase since 2013.[23] 'Single use' refers to products which are made to be used only once before being thrown away.

There is some debate about its definition, particularly, and perhaps unsurprisingly, among the brands that put single-use products on the market.

The final straw

There is no debate that certain items are very difficult to recycle; straws, pouches and black plastic, which cannot be sorted (page 108), are all single-use plastics as they are used once and then thrown away. In fact, straws, stirrers and plastic cotton buds have such an impact on the environment if disposed of incorrectly that they have now been banned in the UK (page 207).

The main issue with certain plastics that have often been considered as single use, such as plastic bottles or rigid salad pots, is that they are highly recyclable but normally sold 'on the go', where recycling bins can be harder to find. In total, 55 per cent of drinks sold in plastic bottles are consumed outside of the home.[24]

A desire for convenience has led to this issue, as consumers want to be able to buy food and drink in easy-to-use containers while on the move. When deposit return schemes are eventually introduced in the UK (page 56), the rate of plastic bottle recycling could rise to over 95 per cent.[25] So if they are collected and recycled back into bottles, is this single use?

The view of big brands is 'no', since if the bottle is recycled, the plastic is used more than once. However, environmental campaigners would argue that the bottle is still single use, as in order to recycle it into a new bottle, further energy and other resources are required. What is produced is essentially a new bottle, different from the old one, and therefore the original bottle has only been used on a single occasion. Refill should therefore be the only way for an item to be classed as anything other than single use.

For good reason, single-use plastics are generally considered to be bad for the environment. However, the biggest issue with a move away from single-use plastic is the unforeseen and unintended consequences: often, single-use plastic can have a helpful and positive purpose. Consider fresh foods in the supermarket. A cucumber wrapped in plastic film will last three days longer than an unwrapped one, which constitutes a crucial time increase in a complex supply chain. Without this type of plastic, food waste would dramatically increase. When we take into account the level of energy required to cultivate, grow, harvest and transport the food to our fridges, the environmental impact of this wastage could far exceed the impact of a small piece of plastic film (which should soon become recyclable, page 233).

To avoid the stigma associated with single-use plastic, there has been a dramatic increase in compostable (page 225) and biodegradable plastics (page 222). The producers of these types of packaging claim that they are not plastics. However, they *are* plastics, simply with a different composition. As such, these sorts of plastic are just as much 'single use' as their traditional counterparts: they are pieces of packaging, made of plastic, that will only be used once. For more detail on the challenges of biodegradable and compostable packaging, see pages 222 and 225.

There is actually another subcategory of plastics, called bioplastics (page 221). These are made from sustainable sources, like plants, rather than fossil fuels, with the word 'bio' referring to the source. These plastics may not be compostable or biodegradable, but their origin can be more environmentally friendly than traditional plastic.

This is a very complex issue, which will be explored throughout this book. Ultimately, it comes down to the convenience demanded by consumers. If, as a society, we are willing to embrace reuse systems, buy local or, better yet, cultivate our own food, then it is possible to eliminate single-use plastics.

However, if we continue to buy from supermarkets with complex supply chains, then we should not demand the elimination of single-use plastic, as this could cause even more environmental harm. We should instead demand that the collection and recycling infrastructure develops to allow this packaging to make its way through the recycling system.

As consumers, we have the power to shape policy and affect the approaches of large retailers via our buying habits. As consumers, it is up to us to make educated decisions on what products we buy and how we choose them.

2

The Rubbish History

Lots of innovation has got us to where we are now. While plastic use skyrockets, we must wonder where it all came from in the first place? All packaging is modern in the landscape of our planet's history, so let's explore some of these timelines a bit further.

The history of paper

Paper has been used as a packaging material for thousands of years. Necessity is the mother of invention, and paper was born out of the need to write, keep records and transport delicate goods.

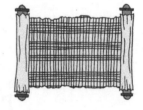

Crease to exist

The word 'paper' is derived from the word 'papyrus', a material used as a precursor to paper. The *Cyperus papyrus* is a plant from which Ancient Egyptians would take thin strips and line them up to make a sheet. Another layer of strips would then be added on top at right angles to the first sheet. The sheets would be bashed together to form a surface that could be written on. In contrast, modern paper is made by breaking down fibres before pressing them back together, producing a much smoother and stronger surface.

Over 2,000 years ago in China, hemp paper was used to wrap and pad delicate mirrors. The inventor of paper is widely considered to be Cai Lun, a court official during the Han Dynasty (202 BC–AD 220), although earlier examples of paper have been found. Over the next thousand years, toilet paper, paper money and bags to contain tea were all used in China.

The fifteenth century saw the growth of the paper industry in Europe, with the creation of printing presses, which transferred ink to paper mechanically. This led to an explosion in the number of paper mills across the continent, as demand for paper increased.

The early nineteenth century brought the development of the Fourdrinier machine, which produced a long roll of paper, rather than individual sheets, marking the advent of modern papermaking. These machines allowed paper to be made from wood pulp, a previously undiscovered technique.

The history of cardboard

There are broadly two types of cardboard: paperboard and corrugated board. Paperboard is thicker than paper; it's the kind of board that makes up a cereal box. Corrugated board is a pleated paper covered by two sheets, and is mostly found in thick cardboard boxes.

Cereal filler

Like paper, cardboard originated in China around the fifteenth century. However, the cardboard box was not introduced until 1817, with early boxes joined together piece by piece. The end of the nineteenth century saw the invention of the pre-cut cardboard box, a single flat piece which folded into a box. The inventor, Robert Gair, came up with the idea accidentally when a metal strip on a press machine he was using was set too high; instead of folding the paper, it cut it, leading him to realise that the machine could simultaneously cut and fold board.

Kellogg's was the first company to use paperboard boxes for its cereal, over a hundred years ago. This dramatically increased the use of cardboard as a packaging material.

Corrugated cardboard was patented in England in 1856 by Albert Jones, to be used as a liner for tall hats to make them more comfortable and durable. It was not until 1871 that corrugated cardboard was patented in box form, the strength of the pleat providing an advantage when transporting goods.

Initially, corrugated cardboard had a pleat on one side and a single liner sheet, and it was used for wrapping glass bottles. Three years later, the design was improved by Oliver Long to include a liner sheet on both sides of the pleat, the same as the corrugated cardboard we use today.

The history of glass

Naturally occurring glass has been traded as a commodity since the Stone Age. Glass made from rapidly cooling volcanic magma was used in tools and weapons and was highly valued.

Ship and a bottle
Visit a museum and you may notice examples of glassmaking

across the centuries and millennia. The first glass objects, mainly beads, date back to around 3500 BC. According to legend, glass was discovered in Syria when merchants were shipwrecked. They traded in natron (a compound found in dry lake beds in Egypt) which they used to prop up their pots while cooking. The natron mixed with the sandy shore and, heated by the fire, created molten glass.

Glass was originally made using moulds. However, the innovation that advanced glass around the world was the discovery of glass-blowing in the first century BC. Around AD 100 manganese dioxide was introduced to the glass mix, creating transparent glass which could be used architecturally.

In England in 1674, George Ravenscroft discovered that adding lead oxide to the molten glass improved its appearance and made it easier to melt. This allowed clear lead glass to be produced on an industrial scale, making England a world leader in glass production.

Only since 1912 has the manufacture of glass jars and bottles been cost-effective enough to be carried out industrially. In 1904, Michael Owens invented an automatic glass bottling machine, which sucked molten glass into a mould to be formed into a bottle or jar. Karl Peiler improved the design in 1915, focusing on delivering just the right amount of molten glass to the machine. These machines were so efficient that in an hour they could produce more bottles than a team of glass-blowers could achieve in a whole day.

Making glass weigh less (lightweighting) has been supercharged by new technology in recent years, which is making glass a more sustainable material. Process improvements essentially allow for a more even spread of glass, reducing the margin for error in glass thickness and allowing for a strong bottle made of thinner material.

The history of aluminium

Aluminium metal is a relatively recent discovery, due to the complex processes required to refine it from its ore. Aluminium is named after an aluminium-based salt, alum, which was discovered and used long before the metal.

Very refined

Aluminium was first extracted from its ore in 1825 by Hans Christian Oersted from Denmark. His work was continued by a German chemist, Friedrich Wöhler, who spent eighteen years experimenting to create small balls of aluminium. Aluminium was difficult to refine and at this time it was more expensive than gold.

The first industrial production of aluminium was created by Henri-Étienne Sainte-Claire Deville in 1856. Its availability was increased significantly by the development of the Hall–Héroult process in 1886: dissolving aluminium oxide in a mineral called cryolite, which halved the aluminium oxide melting point. The molten mixture is then electrolysed to produce aluminium.

Aluminium ore is called 'bauxite', named after the French village Les Baux-de-Provence where it was discovered in 1821. The Bayer process, developed by Austrian chemist Carl Josef

Bayer in 1889, is used to refine bauxite to aluminium oxide for feeding into the Hall–Héroult process. These two methods are still used today to refine aluminium.

Aluminium turned out to be a very useful metal. It is light and corrosion-resistant, and therefore an effective type of packaging. Recycling aluminium has become a necessity due to the energy and complex processes required to create it. Melting used aluminium requires only 5 per cent[1] of the energy used to make aluminium from ore, which is why aluminium packaging has a very high collection and recycling rate in the UK with 82 per cent of aluminium cans and 68 per cent of all aluminium packaging being recycled in 2020.[2]

The history of steel

Unlike the relatively recent discovery of aluminium, steel has been used for 4,000 years. The earliest production of steel dates to around 1800 BC, when it was used in weaponry. The steel industry has played a key part in the economic development of many countries due to its importance in infrastructure.

Steel yourself

The modern steel industry began in the mid-nineteenth century. In 1856, Henry Bessemer made a discovery that led to steel being transformed from an expensive metal, used primarily in armour, knives and swords, to a key element in construction and infrastructure worldwide.

Steel is an alloy of iron and carbon. Iron ore is mined and then smelted into 'pig iron', which has a high carbon content (3.8–4.7 per cent), making it brittle and not particularly useful. Henry Bessemer's innovation was to blow air through the molten pig iron, which reduces the impurities including the high levels of carbon found in iron ore. This created molten steel, which was cheaper and easier to make than previous methods and contained only 0.2–2 per cent carbon.

Throughout the late 1800s, wrought iron rails were replaced with stronger steel, shipbuilding was improved by using this cost-effective metal and by 1875, Britain accounted for 47 per cent of the world production of pig iron.[3] During the twentieth century, furnaces were improved to ultimately replace the Bessemer process, resulting in more predictable steel outputs.

The use of steel as the outer layer of a tin can was first patented in 1810 but not put into practice until three years later, when it was employed in canning food for the Royal Navy. By the late nineteenth century, the public could buy condensed milk, meat and baked beans in tin-lined steel cans, boosting the use of tin cans as packaging.

The history of plastic

Plastic is the most modern of all the packaging materials in this section. Its lightweight nature and multiple forms make it the most complex and prolific material in our product supply chains; around a third of all plastic is used in the packaging industry.

Plastic elastic

Plastic comes from the Latin 'plasticus', which means 'able to be moulded'. The word was in use long before it was attributed to the category of polymers we now think of when we hear the word 'plastic'. Over the last 150 years we have learned to make synthetic polymers from the carbon atoms contained in petroleum, through a process called polymerisation.

The first known human-made plastic was called Parkesine, and was developed by Alexander Parkes in Birmingham in 1856. Parkesine was made from cellulose, which is found in the cell walls of plants, mixed with nitric acid. Once dissolved in alcohol, this mixture would harden into a plastic material that could be moulded when heated.

In 1907, Leo Baekeland invented polyoxybenzyl-methylenglycolanhydride, more commonly known by the snappier name Bakelite. This was the world's first fully synthetic plastic; it contained nothing found in nature. The success of this new plastic led many major chemical companies to invest in the development of new polymers.

Mass production of plastic really developed from the 1920s with the commercialisation of PVC (1926), polystyrene (1931), polyethylene (1933), PET (1941) and polypropylene (1957). These are all examples of thermoplastics, which will soften on heating and harden when cooled.

The history of wooden pallets

Wooden pallets are used to transport goods around the world and store products safely. Every manufacturer, warehouse and factory will use pallets in their supply chains. Around 5 billion pallets are made each year, and

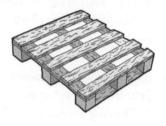

wooden pallets have an average lifespan of around three years.

A pallet cleanse

The predecessor to the wooden pallet was a 'skid', a simple piece of wood which reduced friction, making it easier to move the goods piled on it. Skids can be traced back to Ancient Egypt.

The modern pallet is designed to have two layers, so it can be lifted by a forklift truck. Therefore, modern pallets were born from the development of the forklift truck in 1915. The earliest patent describing the modern pallet is dated 1924, which refers to the pallet as the 'lift truck platform'.[4]

The growth of the wooden pallet was linked to the need to move goods and weapons around during World War II. This led to a significant increase in demand for the pallet and the invention of the four-way pallet, which allowed the pallet to be lifted from all sides.

Pallets have reinvented product transportation, enabling up to 1 tonne of goods to be picked up and moved in a single lift. The relatively simple design has not significantly changed for about a century.

Today, a pallet may be made of plastic as well as wood. With its increased strength, durability and resistance to liquid, a plastic pallet may last twice as long as its wooden counterpart. Plastic pallets can also be used in food-production facilities, surpassing their wooden equivalent.

The history of electricals

Long before Benjamin Franklin and his kite, electric fish had been shocking the Ancient Egyptians, earning the nickname 'Thunderer of the Nile'.

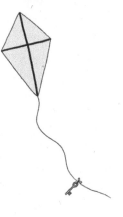

Electricity has historically fascinated scientists, and developments and innovations throughout the eighteenth and nineteenth centuries brought electricity into our homes and workplaces. We now rely on a multitude of electrical gadgets and appliances in our daily lives.

Past to current

It would take a whole book (and then some) to describe the history of electricals up to the latest innovations, so what follows is only a brief summary. Most of it starts with the invention of the diode, a small device developed in 1904 which allowed electricity to move from the cathode to the anode, while controlling the electric current.

The diode became the key to electric circuits and was crucial to developments in radio, television, telephone networks and computers. Since the 1980s, traditional diodes which moved electricity in a vacuum have been replaced by solid-state devices, which are smaller, cheaper and more reliable. Solid state simply means the electricity flow takes place in a solid, rather than a gas, as was the case with vacuum diodes.

The development of circuit boards in the 1960s and 1970s led to the rapid expansion of personal, handheld, low-cost electrical devices. However, this shift to cheaper, replaceable gadgets has also created a throwaway culture with electronics, and a corresponding need to ensure they are captured and recycled carefully.

The history of batteries

Benjamin Franklin first used the term 'battery' in 1749; he was describing a group of 'Leyden jars', which were glass jars filled with liquid and covered with metal on the inside and outside, used to store electricity, essentially the earliest battery.

Pile of rubbish

The first true battery was invented by Alessandro Volta in 1800, using his knowledge that electrical events were caused by two different metals and a liquid agent (an 'electrolyte'). His battery, the 'voltaic pile', was made from discs of copper and zinc, sandwiched around a cloth soaked in brine. Essentially a battery transfers chemical energy into electrical energy, which is what was observed in these early batteries.

The voltaic battery charge was short as hydrogen bubbles would form around the copper discs, increasing the electrical resistance of the battery. This phenomenon was solved by John Daniell in 1836; the Daniell cell used a second electrolyte to remove the hydrogen being created.

Early batteries were 'wet cells', i.e. they used a liquid, which was impractical for day-to-day use due to leakages and the need to keep them orientated a certain way. 'Dry cells' were patented in 1886 by Carl Gassner, a German scientist. His batteries used a paste rather than a liquid as the electrolyte, creating a practical and portable alternative to previous batteries.

Over the course of a battery's life, the chemicals are used up. These are called primary batteries and are non-rechargeable. Rechargeable batteries, invented in 1899, use electricity to reverse the chemical reactions, allowing them to be used again.

The history of textiles

When we think of recycling, clothing is unlikely to be the first thing that springs to mind. However, fashion has developed into big business, and fast fashion is bringing its own environmental challenges. How did we get to a place where clothes are

sold so cheaply that they are also in danger of becoming single use?

Fast fashion

The industrial revolution brought huge changes to the world of clothing, as garment production moved away from being a cottage industry and into factories, where bulk items were manufactured in a range of sizes. Developments like the modern sewing machine in 1846 shifted the focus away from made-to-order towards mass production.

World War II led to restrictions on fabric availability and an increase in standardised clothing, resetting public perception on how clothing could be sold. This made mass production more acceptable after the war and led to a boom in the fashion industry.

The 1960s brought with it new style trends, a burgeoning youth culture and the expansion of shopping centres, which led to the kind of practices we call 'fast fashion': affordable clothing used for a relatively short period of time. Large textile mills and clothing factories opened around the world, particularly in developing countries with cheaper labour costs. The brands and retailers we buy from today expanded rapidly.

Like so many things in material selection, there are many issues that stem from the drive to buy at the lowest cost. Unethical working practices in developing countries, harmful chemicals and huge amounts of waste all contribute to the production of clothing which is used for a relatively short time. However, some consumers are becoming increasingly aware of the true cost of fashion and are altering their buying habits to opt for more ethically produced and longer-lasting clothing. This is also coupled with a rise in apps and websites dedicated to trading second-hand clothing, which is leading to a change in younger consumers' buying and selling habits.

The history of all these products is useful to look at as we move deeper into the world of recycling. This section has demonstrated the length of time it has taken to get to where we are today and the complexity of product development.

While we may hope the world of sustainability can move faster, big business has now built up around these innovations, and change will be slow and methodical. Recycling and sustainability are relatively recent innovations, compared to the length of time these products have been around, and their scale in modern society is unprecedented.

The next sections exploring the collection, sorting and recycling processes will help us work through the merits of these different material types.

3

The Rubbish Collection

Here we are at the start of the waste management process. We are creating rubbish at home and on the go; how does it actually get from us to the recycler? Why is it not a consistent process, and does it really need to be so complicated? Time to find out…

How is waste collected?

Most household and business waste collections are carried out using trucks. Packaging, by its nature, is usually light in weight and therefore cost-effective transport is  achieved by moving high volumes around the country. For waste management companies, the trick to achieving high-quality recycling is to minimise contamination (page 53) by designing collections that separate any liquid and paper. Therefore, some councils will use trucks that have multiple compartments to separate out materials, in order to prevent contamination.

There are three main methods of collection, each with its own advantages and disadvantages. Councils all around the country will collect differently, so you need to check local websites to see how yours carries out its collections.

Multi-stream collection

Multi-stream recycling requires multiple bins, and usually relies on the resident to separate packaging before putting it out for collection. Sometimes this will be through a kerbside sort (page 79). An example of effective multi-stream collection would be to have plastics and metals collected in one bin, card and glass in another bin, which might then be separated when it is put on the truck, and paper in a third bin on its own, to reduce contamination.

Two-stream collection

Two-stream recycling requires two bins and mixes the plastic, metal, cartons and glass; these materials will then go through a Materials Recovery Facility, or MRF (pronounced 'murf'), for sorting (page 66). Paper and cardboard are then usually collected in a second bin, which prevents liquid from bottles or cans ruining the fibrous material. Collecting in this way will also prevent glass shards getting into the paper recycling process and damaging machinery.

Co-mingled collection

With co-mingled recycling, everything is mixed in one bin. This generally leads to the highest level of contamination but is very simple for the household to manage. This stream of recycling also goes on to a MRF for sorting (page 66).

What is a waste carrier?

If a business has its waste collected, it is under an obligation to ensure the person conducting the collections is a registered

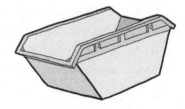

waste carrier. This legal requirement reduces fly-tipping and illegal waste management. Any business that moves waste as part of its normal business activities must be registered as a waste carrier.

Where you bin?

Waste carrier requirements do not apply to householders, which is why you can freely drive your rubbish to a civic amenity site. It also explains why a business cannot do the same as easily. The registration requirement does not apply to businesses moving waste around the same premises.

A key target of the waste carrier requirements are companies that make a profit from moving waste. Without a central register, companies could move waste illegally and dispose of it unsafely without consequences. This legislation attempts to log companies that move waste in an effort to deter this activity. It is not a perfect system as it is still open to abuse, but it does prevent people who have been convicted of crimes related to waste disposal from registering. The requirements also cover brokers and dealers of waste who may not actually handle it, but simply arrange and oversee its transport. If you are having waste collected, the register provides a quick and easy method to verify that the collection company is registered.

There are two tiers of waste carrier: lower and upper. The lower tier includes companies moving the waste they have produced themselves, for example a manufacturing plant moving some production waste to another site. It is free to register with the lower tier. The upper tier is for carriers managing other companies' waste and entails a small registration fee every three years.

Why don't towns and cities collect the same way?

One of the biggest challenges to achieving high-quality and easy-to-understand recycling is that all local authorities collect in a different way, with different bin colours, bin sizes, frequency of collection, materials collected and so on. Across the country there are over 350 different collection methods!

Slow changes, fast news

Recycling in a city is managed either by a waste collection company that is contracted by the local authority to deliver the services, or in-house. Waste management contracts are often very long, as collection companies and recyclers will need to invest in particular types of equipment in order to facilitate the collection and recycling of specific materials. Since each contract will end on a different date, local authorities cannot all move to the same system at the same time.

So many things compound this issue, including decision-makers at the local authority and what their definition of 'good' looks like. Local authorities may also feel they need to be distinct from neighbouring cities, for example, by ensuring they use different coloured bins or collection styles. This is done in the belief that it will stop residents who may live near each other from spreading confusing messages; invariably it will do the opposite.

In addition, local geography plays a part in what is collected. As an example, areas with narrow streets would not be accessible for the usual large collection trucks. A smaller truck is required, which may have fewer compartments to sort waste, leading to a lower quality, co-mingled collection (page 46). As the routes taken by trucks are designed to save fuel, this could affect the whole route, even in areas with wider streets.

Speed of collection is also extremely important in densely populated areas, leading to an abundance of co-mingled collections, which reduces the quality of collected materials.

These issues are compounded by the speed at which public feeling changes towards recycling and types of packaging. For example, a drive towards facilitating the kerbside recycling of flexible plastic film, such as the plastic around fresh fruit and vegetables, will require collection companies to adapt their process to account for an additional new material. The long-term contracts between local authorities and waste collection companies do not easily lend themselves to adapting to changing public opinion.

The future introduction of deposit return schemes (page 56) will transform what councils collect, very likely moving profitable waste (like bottles) away from local authorities and impacting on the economics of collections. This will create more space in the truck and ultimately provide the opportunity to collect a wider selection of materials from the home, e.g. plastic film, small electrical appliances and fabrics.

How does my local authority decide what to collect?

Which materials your local authority chooses to collect for recycling is a question of infrastructure, including the type of recyclers that are local to the area or the local authority's access to national or international recyclers.

Collected ≠ recycled

A waste collection company does not just have the contract to collect waste on behalf of the local authority – it will also decide where the waste gets sent once collected. This has led to some of the questions we hear today around why waste is being shipped out of the UK for recycling.

As transport is a key cost to collection companies, and moving lightweight waste is not particularly cost-effective, they prefer local sorting facilities or recyclers (if the waste is already sorted). This makes the collection company reliant on the capability of sorters nearby. This can pose a problem if, for example, the local sorter is unable to sort different types of plastic tubs for recycling: if that is the case, the local authority will be unlikely to collect them. Or, arguably worse, the material may be collected but then not recycled at all.

This is a key problem within the world of recycling: if you collect some material, but the necessary sorting and recycling infrastructure for it does not exist, is it recycled? How do you communicate this to the consumer who is diligently sorting their waste in the belief it is going to be recycled?

Recycling is often faced with this 'chicken and egg' situation. Recyclers will build infrastructure when waste is collected. Local authorities tend not to collect waste without the recycling capability being in place. Therefore, achieving consistency of collections at the household level is extremely challenging without also looking at the funding and development of recyclers. Future legislation is being designed to tackle this problem by mandating what is collected and providing additional funding to infrastructure development (page 216).

What are specialist collections?

In an ideal world everything would be collected at kerbside; however, decisions about what to collect are still largely driven by commercial value. This means there are obvious collection gaps, which brands are trying to address for their own products by setting up specialist collections.

Post-consumer waste

The commercial drivers to collection mean that once waste falls into the 'difficult to collect' category, such as flexible plastic, it is more likely to require a specialist collection. This could be through the post or via a bin in a supermarket. These collection routes are becoming more popular but do not operate at the same scale as kerbside collection, as they require the consumer to put in extra effort, beyond their normal recycling habits.

These collection methods tend to collect cleaner material, with less contamination, as consumers typically will have to carry them further and therefore are more likely to wash out the packaging and make sure it is correctly sorted.

For supermarket collections, bin placement is essential; bins to collect carrier bags will usually be found at the entrance of a supermarket to prompt shoppers to dispose of their used flexible plastic on their way in. Conversely, bins placed by a checkout are more likely to simply fill up with paper receipts.

On the collection of waste by post, a practice which is becoming increasingly popular, the Royal Mail has been very clear that waste of any kind should not be posted, partly because this would put an obligation on Royal Mail to be registered as a waste carrier (page 47).[1] As such, it is not always clear how these schemes are allowed to work in practice, although arguably without checking each and every package, Royal Mail may not be aware that deliveries contain waste items. It is certainly worth considering this when schemes pop up that allow you to post rubbish. Products like mobile phones can be posted as they are mainly sold to resellers or refurbishment companies.

What is contamination?

Throughout the book we refer to contamination, as it is a key issue in recycling. But what does it mean, and which materials are most likely to get contaminated?

Keen for clean

Contamination means the act of making something unsuitable by contact with something unclean.

This can happen a lot with waste as, by its nature, it is not particularly clean. Generally, this is true of packaging which may have contained food or liquid, although contamination can also occur when materials are simply put in the incorrect bin. For recycling to be effective, it is essential that certain waste types are kept apart in order to retain the quality of the recyclate.

For example, leaving liquid in a bottle without a lid could cause it to splash onto paper or cardboard, which would lower the quality of the recycling. Similarly, grease on a pizza box would cause speckles to appear in the recycled paper it comes into contact with.

Therefore, normally a local authority will require you to keep glass, plastic and metals together, or at least separate from paper and cardboard. This is also why some packaging which contains difficult-to-remove residue (such as coffee pods) is not widely collected or recycled as the chances of it contaminating other waste is high.

All recyclers want contamination kept to a minimum, which is why it is better to split waste streams as much as possible. You can help by always rinsing out your food and drink packaging and allowing it to dry before you put it in the recycling bin.

How is wood collected?

Wood is not usually collected from houses as it is not a common packaging product.

The most common packaging made of wood is pallets used to transport products between business sites, which do not need to be collected at kerbside.

Twig of the dump

Waste wood that arises from the household is generally lower in quality than wood from the construction industry. This is because household wood will often include things like broken furniture which has been painted or glued.

If the wood is ready for disposal then its best chance of getting recycled is to be taken to a local civic amenity site or recycling centre, where there will be a specific collection bin for it.

Household wood is managed in one of two ways: it will be recycled into woodchip or used as a fuel (otherwise known as recovery). Recycled woodchip is used in MDF (a type of wood used to build furniture), although this must not have been painted before being recycled.

Most household wood therefore will end up being used as a fuel. This is because wood burns well and can be used as a source of heat and power. Energy created from burning wood is becoming more and more common and so it is much rarer for wood to be recycled through traditional methods.

An alternative to taking the wood to a recycling centre is to visit a local wood recycling project. These aim to take waste wood from households and businesses and de-nail the pieces, allowing each useful piece a new lease of life via a timber yard or workshop.

What is a deposit return scheme?

Glass and plastic bottles and aluminium cans are very valuable. They are easy to recycle and are in high demand by manufacturers who want to use recycled content in their products. In the UK, most cans and bottles are recycled in kerbside recycling or via public recycling bins. However, in other countries, an alternative system is used, called a deposit return scheme. Voluntary deposit return schemes were common in the UK for decades; however, with the rise of plastic bottles, they were phased out.

Under a deposit return scheme, when a drink bottle or can is bought, it will cost a bit more than the drink would usually be; the extra bit of money constitutes the deposit. When you have finished with the drink, you can take the bottle or can back to any supermarket or shop. Depending on the size of the shop, you can either hand it back to the cashier or put it into a machine called a reverse vending machine and get back the extra money you paid.

A deposit return scheme is set to be introduced across most of the UK in 2024, with Scotland looking to introduce their version earlier in 2022, although these dates could definitely be delayed. More detail can be found on page 216.

Pence makes sense

A deposit return scheme puts additional cost on the consumer, therefore encouraging them to personally ensure their waste is recycled, rather than relying on collections from the household which may be contaminated by other waste.

Deposit return schemes are popular with the public, and over twenty countries are currently using them to subsidise recycling, including Denmark, Australia and Norway. As the UK has focused on kerbside collection, a formal national deposit return scheme covering all bottles and cans has never been introduced. However, this will change soon, as the government has committed to introducing a deposit return scheme. So, do they work?

The short answer is, yes. Norway arguably has the best deposit return scheme in the world and collects over 95 per cent of its plastic bottles, with less than 1 per cent entering the wider environment. Norway has around 4,000 reverse vending machines and 11,000 drop-off points. (By comparison, due to the much higher population in the UK, it is estimated a similar scheme would require over 37,000 reverse vending machines!)[2]

The main concern within the packaging industry about the design of a deposit return system is the unintended consequences, particularly around multipacks of drinks. If a deposit return scheme introduced a deposit of 20p per can, a multipack of twenty-four cans would have a total deposit of £4.80. This would significantly increase the cost of the pack and most likely cause consumers to opt for the same quantity of drink in two-litre plastic bottles, the deposit for which would constitute just 80p.

Some schemes have resolved this by having a different cost for drinks sold in multipacks, but there is concern the UK will not adopt this approach and is instead focusing on a single price per unit, regardless of how it is sold.

Manufacturers have raised concerns around deposit schemes as, while deposits undoubtedly work, bottles and cans are among the items with the highest collection and recycling rates, so deposits are a solution that will only create an incremental improvement. Meanwhile materials with low recycling rates (plastic film and pouches) are excluded from these schemes.

A deposit return scheme will require consumers to bring empty cans and bottles back to shops and supermarkets, requiring a big behaviour change. Consumers in the UK will also have to stop crushing cans and squashing plastic bottles once they have been used, something we have been encouraged to do for the last thirty years – unfortunately, the deposit return systems have a hard time reading barcodes from squashed packaging, so the items will need to be inserted into the return machines intact.

Deposit return schemes are on their way, something that will be celebrated by a significant portion of the population. It will be interesting to see how UK recycling rates improve, considering that we are the first country to introduce them *after* establishing successful kerbside collection.

What are bottle banks?

Bottle banks are recycling bins dedicated to the collection of glass. With the introduction of glass collection from the household, bottle banks are  used much less frequently than they once were, but you can still find them in many supermarket car parks.

BYOB

Glass can be infinitely recycled and turned into new products with a high recycled content. To ensure glass is recycled effectively, it is necessary to ensure the glass was sorted by the correct colour. The first bottle bank was introduced in 1977 to facilitate this and there are still over 50,000 of them in the UK today.

Since glass keeps its colour after it is recycled, bottle banks require the consumer to separate the colours green, brown and clear; any blue glass is collected with the green. Collections from the household may be sorted by colour using machines so do not need to be separated at the kerbside. As the banks are focused only on glass and the bottles are sorted by consumers, bottle banks are the most efficient way of collecting uncontaminated glass.

The colour of glass presents an interesting situation. The UK uses more clear glass in its manufacturing, but imports lots of green glass in the form of wine and beer bottles. As we collect more green glass than we need, green recycled glass is more likely to be exported or used in aggregate applications (page 68).

Glass can be recycled without degrading the quality of the product. Recycling glass is significantly better for the planet than creating new glass: the recycling process uses less energy as glass cullet has a lower melting point than glass manufactured from raw materials. Therefore, bottles should always be recycled, even if it means a trip to your local bottle bank.

Interestingly, the future introduction of a deposit return scheme is bringing bottle banks back, as local authorities move away from collecting glass at kerbside.

How are waste electricals collected?

In January 2007, new laws were introduced to improve the recovery and recycling rates of waste electricals. As we all use more and more gadgets, it has become essential to ensure this waste is collected and recycled or fixed for reuse.

Powerful collections

In 2018, worldwide electrical waste (e-waste) had reportedly reached 50 million tonnes, a figure expected to double by 2050. The vast majority (around 80 per cent) of this ends up in a landfill site.[3] Globally, this is a significant problem, as most uncontrolled e-waste is shipped to developing countries and dealt with by some of the planet's poorest people. This normally involves burning the waste, which is harmful both to the individual's health and to the wider environment.

Therefore, the key thing with waste electricals is making sure they do not end up in landfill: anything that has a plug or needs a battery should be separated and not put in the normal rubbish bin.

There are a number of ways to ensure your waste electricals are collected. You can take them to a local recycling centre where there will be a special location for them, or arrange for the council to come and collect them, though this will probably involve a fee.

The easiest way to get your waste electricals to the right place is to give them back to the shop when you buy a new product. A recent change in the law means all major retailers will have to take back your electrical waste when you buy a new product. This is on a like-for-like basis; for example, if you buy a new kettle, you can return your old one at the same time, or if the purchase is large (like a fridge), the retailer should come and collect it from your house. The recycling service will always be free of charge, although the retailer can charge the transport costs. It is available even if you did not buy the old product from the same shop as the new one.

How are batteries collected?

In February 2010, new laws were introduced to improve the collection rates of small batteries, which were frankly a bit rubbish! Since 2010, battery recycling has increased from 9 to 45 per cent.[4]

Batteries contain valuable metals and poisonous chemicals, but as they are so small it is easy for people to chuck them in the bin. It is important you ensure this does not happen.

Fully charged containers

The law requires any shop which sells a minimum of four packs of batteries a day (so not very many) to ensure they have an easy-to-find battery collection point. Since 2010, collection points have sprung up everywhere, in most shops, and even in schools and offices. They normally look like buckets, tall tubes or plastic barrels. Experience shows some stores do not clearly display their collection points, so you may have to look for the bins.

Some local authorities will also collect batteries from the kerbside if they are in clear plastic bags. However, this is not as common, so it is important to check local websites for information.

There is no legal target on recycling car or industrial batteries, as they are regularly recycled due to their size and value. In fact, it appears that some of these batteries are making their way into the reported 45 per cent recycling rate, as lead acid batteries (page 76) make up the bulk of recycling currently, meaning consumer battery recycling is much lower in reality.[5]

The relatively recent increase in technology that requires batteries has led to a rise in battery fires in rubbish trucks and waste sites (page 199). This is because when batteries are crushed or they touch metallic objects, they can short-circuit, which can start a fire. This is another important incentive to ensure they are collected safely.

What happens after collection?

All three collection routes (page 46) require an element of sorting at the kerbside, the recycling centre or, in the case of co-mingled waste, a sorting plant called a MRF (page 66).

Too picky

Most of the sorting process is automated. However, before the machine-led process starts, a manual sorting process often takes place: the waste is loaded onto conveyor belts and travels along a line to a team of people who manually pick out any obvious contaminants such as crisp packets, plastic bags, toothpaste tubes and the odd Persian rug. This is called 'non-target material', i.e. material not suitable for a future recycling process. In the future, there may be other ways to recycle some of these difficult-to-recycle streams (page 233).

Picking manually reduces the risk of machines getting clogged or damaged by items that should not be involved in the process, such as plastic bags getting tangled in the machinery.

The sorting process is dictated by what a MRF is designed to accept and support, which is why it's difficult to make changes to what a council collects (page 49). Different recyclers that will ultimately receive the sorted waste have different quality standards and accept varying levels of contamination, so this can also determine the level of sorting required.

Sorting waste is an interesting process involving cameras, near infrared technology, magnets, lots of air and giant, rotating sieves. In the next section, we will explore each material in turn and explain how they are sorted.

4

The Rubbish Sorting

Our waste has been collected, it is all mixed up and needs to be sorted in order to be valuable to a recycler. Assuming the sorting hasn't taken place as part of the collection, it is time for our waste to head to a MRF (Materials Recovery Facility, pronounced 'murf').

What is a MRF?

This chapter will explore the sorting process for each material, assuming waste is collected co-mingled (page 46). If waste is collected in this way, it needs sorting and will likely go to a MRF.

Sorting out problems

There are around eighty-five large MRFs currently accredited to receive waste in the UK. They are designed to separate co-mingled recyclables into individual materials before they are sold. Separating different materials increases the value of the waste as the separate streams are easier to recycle further down the chain.

In an ideal world, materials would be collated and stored under cover. Bad weather, and the water that comes with it, can reduce the value of materials, particularly paper and cardboard. In practice, MRFs will vary significantly in setup depending on size and location.

Prior to automated sorting, a MRF will normally involve a 'pre-sort'. This is the manual sorting process to remove obvious contaminants such as glass, wood and wire which could damage the MRF equipment.

The MRFs described in this chapter assume recyclate has been separated from household waste. There is another type of MRF, known as a 'dirty MRF', where recycling that is still mixed in with household rubbish is taken. This can be sorted but will ultimately create lower quality recycling as it will be contaminated with other waste. A typical recovery rate from a dirty MRF can range from as low as 5 per cent up to 45 per cent,[1] showing how variable the waste received can be. For this reason dirty MRFs are not used commonly today as they are rarely economically viable to run.

How are paper and cardboard sorted?

One of the first jobs in a MRF is to separate the paper and cardboard. A common method is to use a giant sieve, called a trommel (the German word for 'drum'), which separates the large 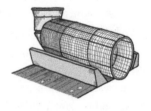 paper and cardboard from the smaller plastics, glass and metals.

Light and airy

A trommel is a steel drum with holes, just like the inside of a washing machine but much longer. Like your washing machine, it rotates, which transports the material being put in from one end to the other. Plastic, glass and metal fall through the holes, whereas the paper and cardboard are too big, and remain inside.

This generates a clean stream of paper and cardboard at the other end of the trommel.

As paper and cardboard are very light, they can also be sorted using jets of air. The jets are positioned to send the paper and cardboard to another conveyor belt, ready to be piled up and sent on to the recycler.

Cameras can also be used to identify materials before being sorted by air. However, as they rely on identifying the colour of the material, cameras are not effective at sorting paper and cardboard as these change colour when they get wet.

MRFs will typically sort three grades of paper: OCC (old corrugated cardboard); newspapers, magazines and pamphlets; and mixed paper.

The UK exports around 70 per cent of its paper and cardboard for recycling.[2] Exporting has become harder in recent years as countries such as China have focused on their domestic recycling, rather than importing waste. These traditional export destinations have either reduced the volume of paper and cardboard they will take or increased their quality requirements. As a result, early separation at a MRF, which reduces the likelihood of contamination, has become key.

How is glass sorted?

Glass jars and bottles are sorted and recycled in the same way. Ideally, they would be colour-sorted via a bottle bank (page 58). The glass that is collected co-mingled needs to be separated from other waste at a sorting facility, first by material and then by colour.

Cullet and colour

MRFs tend not to accept glass, as it can damage machinery. This is why bottle banks became so popular and why today, most local authorities require glass to be separated from other materials by householders, or for collection crews to sort it at the kerbside before putting it onto the truck. Once collected, glass will typically be sorted in a specialist facility.

The weight of glass does make it a very easy material to separate from other, lighter materials. When glass is crushed it will break into small shards, known as cullet. The process of crushing glass will remove any caps or lids left on the jars and bottles (page 100), which are usually picked out by magnets. Hot air is then blown over the glass to dry it and remove any glass dust and paper labels.

Cameras and jets of air can be used to sort the glass by colour, separating clear, brown and green glass. The glass passes underneath cameras which scan the cullet to detect the colour of the glass. Different strengths of air jet are then used to push the glass into different sections. This glass is called 'remelt' and is generally used to create new jars or bottles.

Lower quality glass can be used as 'aggregate', which means it will be crushed further and used in road building. If the glass is destined to be aggregate, it does not need to be sorted by colour.

How are metals sorted?

Two metals are used in the packaging industry: aluminium and steel. It can be difficult to tell which is which. Luckily, your collection is normally just described as metal, so you do not need to identify the difference.

However, they do need to be sorted before recycling; failing to separate aluminium and steel would affect the strength of the recycled metal, rendering the recycled product useless.

Attractive sorting

Magnets have been used to separate metals since World War II, when electromagnets were used to pull metal out of shredded vehicles.

Aluminium is not magnetic, whereas steel is. If you have a magnet in your house, try running it first over a drinks can and then over a tin of soup or beans. The drinks can should not be attracted to the magnet, as these are normally made of aluminium, whereas the food tin likely will be attracted to it, as these are usually made of steel.

There are two types of magnet used to separate steel: an overhead magnet and a magnetic head pulley. The overhead magnet acts exactly as it sounds: it runs above a conveyor belt, pulling up the magnetic steel and leaving behind any non-magnetic material. The alternative pulley system relies on a magnet within the conveyor belt itself. When the steel reaches the edge of the conveyor belt it remains stuck to the underside of the belt, while other waste and contaminants fall off the end.

Aluminium cans are also separated with the aid of magnets, but, as they are not magnetic, a different technique is required. 'Eddy current separators' are fast-spinning magnets which repel any non-magnetic metal on the belt. This causes aluminium to jump across a gap into a container, while everything else falls onto another conveyor belt.

How is plastic sorted?

When it comes to packaging, non-plastic materials are generally manufactured in a single form: cardboard boxes, glass bottles and metal cans. Plastic packaging comes in many different forms, however, which makes it harder to sort. Plastic is marked with a 'resin identification code' ranging from 1 to 7 (page 122). This code could help the recycler to separate the various types, although in practice that is quite difficult. Mixing plastic resins would produce a low-quality output, so it is important to group plastics based on their properties. Before reading this section, it may be worth familiarising yourself with the seven main types of plastic found on page 122.

NIR, far, whatever you are

Although plastic comes in many forms, it can usually be classified as either rigid or flexible. A rigid plastic is strong, like a bottle, tray or yoghurt pot. Flexible plastic includes crisp packets, plastic bags or anything you can scrunch up. At the moment, MRFs only look for rigid plastics, as these are easier to recycle, although new technologies to recycle flexibles are being developed (page 233).

Most plastic is sorted using 'NIR' (near infrared) technology. An infrared beam is shone onto the piece of plastic, which is then identified based on the wavelength that is reflected. Each plastic resin has a unique wavelength signature. Once the automated equipment has detected the plastic type, air jets blow the plastic into different areas, thereby sorting it. The infrared light will not distinguish colour, so once the plastic has been sorted by type, cameras can be used to separate the different colours.

There are two main challenges to the NIR process. Black plastic has often been dubbed 'unrecyclable' because it absorbs the infrared light and therefore does not generate a signature (page 108). Another problem occurs if a bottle is mostly wrapped in a plastic sleeve, often made from a different plastic type, to prevent inks leaking from the label to the bottle. The NIR scanner may detect the plastic of the sleeve instead of the bottle itself. The solution to this problem is relatively straightforward; producers can use a sleeve that covers less than 60 per cent of the bottle.

The NIR process is not perfect, as some plastic can be missed or sorted incorrectly. There are several alternative methods for sorting plastic gaining traction. One option involves printing invisible barcodes all over plastic when it is manufactured, which could then be read at the MRF, significantly increasing the accuracy of plastic sorting (page 235).

A sinking feeling

Once plastic has been sorted at a MRF, it might go on to a specialist plastic sorting facility, called a PRF (plastic recycling facility), which will separate different plastics by chemistry. One technique used by a PRF is a sink-float tank. These are used to separate different types of plastics or contaminants away from the material. This process can be used at various stages of the recycling process, but the principles are always the same: different plastics and contaminants will float or sink depending on their density.

The sink-float process starts with a tank full of water. Additives may be included to change the density of the water, ensuring a specific plastic is targeted. The plastic is added to the tank; high-density material will sink, and low-density material will float. Once the plastics are separated, they can be recovered and sent on for further processing.

An example of this working well is with PET drink bottles, which will sink, while bottle wraps and lids will float. This process ensures streams sent on for recycling are as clean as they can be.

What are the different categories of waste electricals?

There are fourteen categories of waste electricals. Each has their own recovery and recycling rate, which changes each year. Typically, waste electricals will be received by a recycler from a mixed collection

facility, such as a local civic amenity site. Given the variety of goods, sorting waste electricals would be a huge challenge. For this reason, waste electrical items are not sorted in the same way as paper, glass and plastic. Instead, in order to help identify the make-up of mixed electricals, the government has published protocols which estimate the category breakdown of mixed WEEE (Waste Electrical and Electronic Equipment). The fourteen categories are:

Large household appliances
Larger than a microwave: e.g. washing machines, electric ovens, underfloor heating. Does not include fridges (see cooling equipment).

Small household appliances
Smaller than a microwave: e.g. electric showers, toasters, irons and... microwaves!

IT and telecommunications equipment
Computers and phones, printers, laptops, calculators. Does not include screens and monitors (see display equipment).

Consumer equipment
Technology items that do not fit neatly into the IT category: e.g. baby monitors, radios, set-top boxes. Does not include televisions (see display equipment).

Lighting equipment
Most lights: e.g. ceiling, wall, outdoor lights. Does not include

lamps that use gas, such as neon, and LEDs (see gas discharge lamps and LEDs).

Electronic tools

Drills, saws, sewing machines; this category also includes things like pumps for garden ponds.

Toys, leisure and sports equipment

A lot goes into this category, other than the obvious, such as electric bikes, e-cigarettes and games consoles.

Medical devices

Things that a hospital may use: e.g. X-ray machines, ventilators, dialysis machines. It also includes personal items like hearing aids.

Monitoring instruments

Products used to monitor or control things: e.g. smoke detectors, smart meters, car park barriers.

Automatic dispensers

Devices that distribute products or money automatically: e.g. ATMs, photo booths, vending machines.

Display equipment

Anything visual: e.g. computer monitors, TV screens, advertising screens.

Cooling equipment

Anything that cools things down: e.g. fridges, freezers.

Gas discharge lamps and LEDs

Lights that do not belong in the lighting category: e.g. lasers used in the medical industry, gas bulbs and LED filament lamps, including smart bulbs.

Photovoltaic cells

Solar panels and only solar panels; any other device powered by solar goes into its respective category, for example a solar-powered calculator would go in IT.

Most electricals will arrive at a civic amenity site mixed and will be recycled without being separated, although there are specialist recyclers. These categories exist purely for recyclers to estimate what a mixed load of waste electricals is likely to be made up of.

How are batteries sorted?

Batteries that have been collected in containers (page 61) will be mixed and need sorting into different types according to their chemistry. Sorting batteries into their different chemistries allows for more of the original material to be recovered to make new products.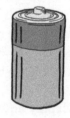

There are numerous methods to sort batteries, which rely on their different shapes, sizes, weights and metals.

Manual sorting is normally first: the batteries run along a conveyor belt with people either side picking out specific types of batteries. Starting with this method also means any rubbish that should not be included with the batteries can be removed. Manual sorting will normally pick out large batteries, such as those found in laptops or power tools.

Magnets will sort batteries by the metal contained in the outer casing. This ensures batteries that are cased in non-magnetic metal like zinc and some button cells are separated from the magnetic types that are likely to contain different chemistries and metals. Most household batteries have steel cases and so are magnetic.

The magnetic batteries are sent for mechanical sorting. This is done by a sieve with different size gaps, like a coin sorter, which shakes, allowing the various batteries to be sorted according to shape and size: button cell, 9 volt, 4.5 volt, AA and AAA, C and D batteries.

As set out above, batteries are made from many different chemistries such as lithium ion, zinc or nickel cadmium. Here is a list of the most common battery types and the different ways the recovered materials can be used.

Lithium ion

Rechargeable batteries, found in phones, laptops and cameras. They can be recharged hundreds of times. The recovered materials include iron, copper, nickel and cobalt.

Nickel cadmium

Rechargeable batteries that were once very popular in portable power tools, torches, cameras and portable electronic devices. They have largely been replaced by nickel metal-hydride and lithium ion; in fact, in the UK they have been restricted since 2009, due to the toxicity of the metal cadmium. Nickel cadmium batteries may only be sold in emergency systems lighting or certain medical applications. Nickel, steel and cadmium can be recovered and used in the metal plating or steel industries; they are unlikely to be recycled back into batteries due to their restricted use.

Nickel metal-hydride

Following the restrictions on nickel cadmium, nickel metal-hydride replaced the technology for portable consumer products like cameras and small rechargeable batteries. They were used in early hybrid cars, but have now been largely replaced by lithium ion as the costs of the technology have reduced. Nickel and steel can be recovered from these batteries.

Alkaline batteries

Sixty per cent of batteries sold in the UK are alkaline batteries – these are the common household types like AA, AAA, C and D.[3] They are not designed to be recharged, and attempting to do so could cause leaks and would affect the performance of the battery. Recovered materials include zinc, manganese and steel.

Lead acid

Most lead acid batteries are used in cars for lighting and ignition. They are also used in things like golf carts. Due to the value of the batteries, they are a recycling success story, with a large proportion of batteries being recycled. The materials recovered include lead and the plastic casing.

What is kerbside sorting?

Recycling collection methods range from multi-stream to co-mingled (page 46). Some local authorities undertake a 'kerbside sort' to improve the quality of the material, ensure contamination is reduced and maximise the material value.

Kerb your enthusiasm

How to collect waste in a local authority is a careful balance of time versus value. How much sorting should the household be required to do? How much should the crew do? Collecting everything at once will require waste to be sent to a MRF (page 66). Requiring the household to do a degree of sorting themselves can lead to 'wishcycling' (page 196). To combat this, some local authorities carry out a kerbside sort, where the collection crew sort the mixed recycling into separate compartments of the truck, allowing them to spot unintended contaminants.

If your material streams are put out for recycling mixed up – for example, paper and glass together, or metals and plastic – it is worth watching how this is collected; do the collection crew keep it mixed on the truck, so that it will need to be sorted later on, or do they divide it between separate compartments when loading onto the truck?

Kerbside sorting is efficient and reduces the time that the waste is mixed for, which in turn reduces the chance of contamination and allows the crew to make sure they only collect things that can be recycled. However, its direct cost is high as it requires more time per household collection. This is offset by the future value of the material and fees that would have been charged by a MRF. This is a careful economic balance weighed up by each local authority.

In terms of quality, material from kerbside sorting is rejected by a recycler less than 1 per cent of the time, compared to an average of 5–15 per cent with a mixed collection.[4]

What happens after packaging is sorted?

After sorting, the recycling is ready to be sold to the recycler, who will turn it back into material for the manufacturer to use for new products or packaging.

Transporting sorted packaging can be very expensive and inefficient, due to its lightweight nature. Transporting air is something all sorters and recyclers try to avoid. Squashing and compacting sorted waste into cubes, called bales, is generally more cost-effective than transporting and selling it loose.

Bale out

The machines used to compact the waste into bales, predictably called balers, could be considered the most important part of the process at a MRF. If there is only one baler and it stops working, the whole sorting line will have to stop while it is repaired.

A baler is usually a large metallic box: when it's turned on, a large plate of metal pushes down on the waste from the top of the box. Once the waste has been pushed fully together, it is squeezed as tight as possible and tied up with wire or plastic banding.

Balers will be selected based on the density, size and weight of the bales they produce. The MRF needs to make sure the bales that it generates match the requirements of the market in terms of size, weight and the material used to tie them up.

Materials that are usually baled include cans, tins and plastic bottles. Paper and cardboard can be sold either loose or baled, depending on market conditions. Glass, which has normally been crushed as part of the sorting process, is obviously not suitable for baling so is sold loose.

Our material is sorted and baled, ready for recycling. Let's now go through the process and explore how we actually transform this waste into something useful and valuable.

5

The Rubbish Recycling

We have reached the point in the waste chain where our rubbish has been sorted into its individual components; now it is time to recycle it! In this section we will explore the actual recycling process for a range of materials: paper and card, glass, metal, plastic, electricals and batteries.

How are paper and cardboard recycled?

After sorting, the paper is ready to be recycled. At this point it has been sorted into types and grades of paper, ready to become new paper.

Typically, paper mills will accept one of three grades of paper: old corrugated cardboard (OCC), newspapers, magazines and pamphlets, and mixed paper, depending on the fibre product they are producing.

They see me rolling

The paper is shredded, breaking it down into very small pieces. These pieces are mixed with a large amount of water and chemicals, causing the fibres that make up the paper to break down. This mixture is called 'pulp' and has the consistency of porridge. The pulp is passed through screens with holes in (like a sieve) to remove any contaminants, for example staples, paperclips, tape and any residual plastics that may have been left on the paper.

The pulp is moved to a tank, where ink is removed using chemicals and air. At this stage, different chemicals can be added to create different types of paper, for example, bleach to create white paper or a thickener to make card. Unbelievably, the pulp is now 99 per cent water and only 1 per cent fibre!

Once the pulp is clean and de-inked it is sprayed onto a conveyor belt. The paper is then pressed by a large roller, which removes more of the water. Finally, the mostly dry pulp passes through heated rollers and is rolled into large reels of paper, ready for distribution.

Unfortunately, each time paper is recycled its fibres are shortened. This means paper can be recycled around six times before it has degraded too much to be recycled any further. Things like paper tissue are an end-of-life paper product, which is why it cannot be recycled (page 184).

How is glass recycled?

Glass is a great material for recycling as, unlike paper, if it is sorted correctly, it can be recycled repeatedly, without losing quality.

New glass is made from sand, soda ash and limestone. These raw materials are all mined from quarries, which have a detrimental impact on the environment, and require more energy to extract them, compared to the recycling process. Each tonne of recycled glass added to the manufacturing process saves 1.2 tonnes of raw material.[1]

Glass half full

After sorting, the glass is normally crushed to the size of a 50p piece. This glass is called 'cullet' and may have been sorted into different colours. Around 70 per cent of the glass that is recycled in the UK is used to make new bottles and jars through a re-melting process. The remaining 30 per cent lower quality glass is crushed to resemble a fine sand and used to make roads or concrete.[2]

Cullet from waste glass is added to new glass ingredients to ensure the output quality is high. Interestingly (and helpfully), the more cullet that is added at this point, the lower the temperature required to melt the glass. It is easier to melt cullet than the new glass ingredients, reducing the energy required to make the glass. The mix is melted down in a furnace at a heat of over 1,500°C.

The liquid glass is separated into globules, which are then blown into new jars and bottles. Ensuring high quality at the sorting stage is essential to make the recycled glass strong and clear enough for use.

Recycling glass is much better for the environment than creating it from new materials. However, it is also a heavy material which must usually be transported reasonably long distances for recycling. Therefore, environmentally speaking, the best thing to do is to wash and reuse glass items at home wherever possible.

How are metals recycled?

By the time metal gets to the recycling process it has been separated into steel and aluminium using magnets and eddy currents. Like glass, metal can be recycled over and over again, without losing quality.

Recycled metal will end up as large blocks or sheets ready to be used to make new packaging, or even trucks, cars, planes and ships.

Iron in the fire

Once metal has been sorted into aluminium and steel, it is shredded into tiny pieces. This gives the metal a larger surface-to-volume ratio, which makes it easier to melt. The metal is then placed in a large furnace, running at the temperatures required to melt the different metals. The melting process can take minutes or hours depending on the size of the furnace. As with glass, the energy required to melt recycled metals is lower than when creating metal from raw materials.

Metal from raw materials is mined from the ground and transported long distances. For example, aluminium comes from bauxite (page 6), and 4 to 5 tonnes of bauxite are needed to produce just 1 tonne of aluminium metal. The amount of energy saved using recycled metals compared to virgin ore is massive, as high as 95 per cent for aluminium,[3] and 74 per cent for steel.[4] It is therefore essential that metals are captured and recycled to reduce the environmental impact of this very versatile packaging.

After melting, the metals undergo specific processes using magnets and electrolysis to remove any remaining contaminants. The clean liquid metal is then carried on a conveyor belt to be cooled down and solidified into the required shapes and sizes. These can be very large bars or sheets ready to be transported and used again in the manufacturing process.

How is plastic recycled?

There are seven main types of plastic (page 122). If the plastic has come from a MRF or a PRF it may have already been sorted. Alternatively, recyclers may also sort plastics by type if they have come straight to the recycler after kerbside sorting. Unfortunately, due to the different uses of plastic, it is more difficult to recycle than other materials which tend to come in a single form, and therefore effective sorting is essential. There are two types of plastic recycling: mechanical and chemical.

Mechanical recycling

The first stage in mechanical recycling is size reduction using a shredder or granulator. A shredder will shred plastic once as it passes through, leading to bigger pieces of plastic, whereas a granulator will keep chopping the plastic up until it passes through holes in a screen. This creates smaller and more uniform pieces called 'regrinds'.

The pieces of plastic are washed to remove any contaminants like dirt, glue and paper. These impurities come off relatively easily once the plastic has been made smaller. Sink-float tanks (page 71) may be used at this point to separate plastic types and any further contamination.

The final step in the plastic recycling process is to turn the sorted and clean regrinds into 'pellets' via an extrusion process. Converting plastic to pellets makes it easier for plastic to be transported and sold to manufacturers. Extruders move the plastic from a hopper down a long barrel using a giant screw which pushes plastic from one end to the other. The long cylinder has heaters placed along it which melt the plastic as it moves, making it more malleable. At the end of the barrel there is a hole called a 'die' which the plastic is pushed through. On the other side of the die, the plastic is cut into small pellets and then left to harden.

Chemical recycling

Chemical recycling is the process of turning plastic back into its building blocks, which can be used to make new plastic. Compared to mechanical recycling, this is a relatively new technology, and lots of companies are currently starting up and working fast to make it a success. Chemical recycling is seen as the solution to a lot of the issues with recycling plastic as it can recycle plastics that have historically been deemed hard to recycle for financial reasons, like flexible plastic. It can be an expensive and more labour-intensive process than mechanical recycling so is likely to complement mechanical recycling, rather than replace it.

There are three main types of chemical recycling: purification, depolymerisation and feedstock recycling.

Purification involves mixing the plastic with a solvent to remove any additives that may have been added to the plastic. The plastic dissolves in the solvent and then, through a purification process, is separated from any non-target material. This process will ultimately create a pure polymer which can be separated from the solvent and additives.

In contrast, both depolymerisation and feedstock recycling deal with the long hydrocarbon chains that make up plastic.

Plastic is made by polymerisation, a process that links monomer chains together. Depolymerisation, therefore, is the opposite process, and will reduce the complex polymers that make up plastic down to a monomer. This process can only be used on specific types of plastic, such as PET, and it cannot be used on PVC, PE or PP. As such, its use is fairly limited.

Feedstock recycling uses heat to break down the polymer chains. The main development in this area is a process called 'pyrolysis'. This involves heating the plastic without oxygen, a method known as 'cracking'. This will create a hydrocarbon vapour which, when cooled rapidly, will produce oils. These oils can then be refined back into plastics or used as a fuel. This process works well with polyolefins, like PP or PE.

The development of chemical recycling has brought some very exciting changes to the plastic recycling market and is explored further later in the book (page 219).

How are waste electricals recycled?

Over 170 million new electrical items are purchased each year, a number that's ever-increasing as consumers upgrade devices more and more regularly in response to rapidly evolving technology.[5] Old electricals, particularly smaller items, tend to sit in drawers and cupboards. However, recycling electricals that cannot be reused or sold is important as they are full of valuable materials.

Shocking developments

The general term 'electricals' covers a lot of products: large household appliances (electric ovens, fridges, washing machines), small household items (toasters, kettles, microwaves), TVs, tools, toys, medical devices, even solar panels! Helpfully, electricals have a crossed-out wheelie bin symbol printed on them to indicate they need to be disposed of correctly when they are no longer required (page 129).

With so much variety, recycling electricals can be complicated and dangerous. Some electricals contain hazardous materials like arsenic, cadmium and mercury. For this reason, waste electricals should be sent for recycling at an authorised treatment facility (ATF) to ensure the recycling is carried out properly and by experienced and regulated companies.

Recovery and recycling rates are set by category (page 73) and measured by weight. This can make things difficult with electricals that tend to weigh less over time as products develop, like televisions: lighter, newer TVs make the statistics appear as though less is being sold, whereas heavier old TVs make it seem that more is being recycled. The recycling targets must constantly be adjusted to balance the changing product mix.

The priority when handling waste electricals is the repair and reuse of whole appliances. Assuming this cannot take place, waste electricals will either be shredded or manually disassembled, depending on what the product is.

Once in small pieces, the process is comparable to how packaging is sorted and recycled. Magnets and eddy currents are used to remove steel and aluminium (page 70). Plastics are sorted using NIR scanners and manual sorting (page 71). This sorting and recycling ensures valuable component parts of the electrical items are captured. Amazingly, one iron will provide enough steel to make thirteen steel cans.

Electricals often contain precious metals which conduct electricity well, making devices smaller and more efficient. Take the smartphone in your pocket; it contains gold, silver, copper, platinum, aluminium and palladium. Remarkably, 1 tonne of iPhones would provide 300 times more gold than 1 tonne of gold ore![6] Realistically, though, it would not be wise to start breaking up your phone to get your hands on some gold as it is present in very small quantities. However, when you consider that over 6 billion people worldwide own a smartphone,[7] the total quantity of precious metals is clearly very significant. It is essential that these materials are reused or recycled and not left to gather dust in a drawer.

It does not end with smartphones; many other electrical items contain valuable materials that can be recycled into something else. For example, gold found in games consoles can be turned into jewellery, zinc found in mobile phones can be used in ships, and plastic from lawn mowers can be used in musical instruments.

How are batteries recycled?

It is estimated that over 178 million batteries are just sitting in drawers in the UK.[8] Recycling them prevents them becoming a fire risk on rubbish collection vehicles (page 199) and ensures the metals they contain are reused.

Batteries are made from different chemistries, so are sorted before recycling (page 76). Once the batteries are sorted, the main aim of recycling them is to recover the raw material used to make the battery so that it can be used again.

Positive story

Lead acid batteries used in cars paved the way for battery recycling. Commercially, it made sense to recover and recycle these batteries as 70 per cent of their volume is reusable lead. However, with the rise in popularity of more volatile alternatives such as lithium ion batteries, particularly in electric vehicles, there is a risk these batteries will be mixed in with lead acid which can lead to fires.

Lead acid batteries are crushed in order to recover sulphuric acid, the plastic casing and metal. The lead is melted down to make new batteries.

Alkaline batteries, which are the common household types, are shredded to separate the paper, plastics and metal. The plastics and paper will be used for energy recovery; they are incinerated to produce heat, which powers the plant (page 197). The metal is melted down to be used in new batteries.

Once the mechanical shredding process is complete, a substance called 'black mass' is left over. This material is the part of a battery that is currently difficult to recycle and looks like a very dark powder. The black mass contains the inside of the battery: the electrolyte, zinc, manganese oxides and other metals. This material can be chemically recycled to recover the zinc and manganese, which can be reused in the steel industry.

So now we have made it through the complete process: collection, sorting and recycling. Our waste has been turned into something new. However, there is some really important knowledge for us still to learn as well as useful, easy-to-remember tricks when it comes to recycling. Read on to build your Rubbish Knowledge.

6

The Rubbish Knowledge

The information contained in this section is some of the most useful and interesting. It will help round off your knowledge of the world of waste, taking you beyond the process and into more practical application.

What is the waste hierarchy?

The waste hierarchy is a simple tool to summarise the order of waste management methods, from least favoured to most favoured. The concept of a waste hierarchy was introduced in 1975 in the EU's Waste Framework Directive, which focused primarily on waste minimisation.

Initial proposals and designs lumped things like recycling and recovery (incineration) together. This was flawed, as recycling is generally less carbon-intensive than incineration (page 197). The six-step hierarchy we use today was introduced by EU directives in 2008 and implemented in UK law through the Waste Regulations 2011.

Pyramid scheme

The waste hierarchy is usually presented as an upside-down pyramid, in which each section represents stages of production (prevention and minimisation) and waste management (reuse, recycling, energy recovery), before disposal is considered, with area size representing desirability (the bigger the proportion of the pyramid = the more desirable).

Below is a summary of each section of the pyramid from the most favoured way of dealing with waste to the least.

Prevention

Preventing the product that becomes waste from being placed in the shop or purchased in the first place is by far the most favoured way of dealing with waste. This is an area that has attracted considerable focus, particularly from members of the public targeting a plastic-free lifestyle. Examples of waste being prevented include buying vegetables with no plastic wrapping, not using a straw, or using reusable cups for your daily coffee (which is both reuse of the cup and prevention of a new one).

Minimisation

Reducing the amount of packaging around a product. Most people are not aware that it is the law to minimise the packaging required for a product. Packaging minimisation is covered by a piece of legislation called the 'Packaging (Essential Requirements) Regulations' (page 14). This regulation requires companies placing packaging on the market to minimise its weight and volume to the level of customer acceptance. So, the next time you receive a small product delivered in an oversized box, you can contact the trading standards department of your local authority and let them know.

Reuse

Reusing something that could have been waste, for example refilling a coffee jar with pencils or reusing water bottles when they are empty by filling them with tap water. This is better than recycling as it uses raw materials in new ways, without expending the energy needed to recycle them.

Recycling

Turning waste into a new product. It would be preferable if the concept of recycling in the waste hierarchy was split in two, as there is a difference between an aluminium can or glass bottle that can continually be recycled without a loss of quality, compared to plastic or paper which will degrade each time it is recycled.

Energy recovery

Recovery is the term given to the burning of waste (page 197). In theory, all rubbish contains energy which can be released when it is burned. This can create heat and it is usually the easiest way to deal with waste if it is not going to landfill.

Disposal

Simply putting your waste in the normal rubbish bin – the least preferable way of dealing with waste. Disposal involves no recycling or energy recovery, and the waste is simply sent straight to a landfill (page 195) to decompose over time.

What is reduce, reuse, recycle?

Is this just a different way of talking about the waste hierarchy (page 96)? Sort of. However, 'reduce, reuse, recycle' (also referred to as the 'three Rs') came first.

The history of the three Rs is debated, but it is believed to have been first coined in America, as a result of the first Earth Day on 22 April 1970. This event saw over 20 million Americans come together in an effort to understand the impact their disposable culture was having on the planet. The first Earth Day actually led to the creation of the US Environmental Protection Agency: the show of strength from the public led to politicians agreeing to set up the agency.

Reduce, reuse, recycle is an easy-to-remember rule for how to approach our consumption of products and how to dispose of them at the end of their life. The three Rs are written in the order you should consider them.

Reduce

Reduce means creating less waste in the first place, similar to minimisation and prevention in the waste hierarchy. This means choosing products that have smaller amounts of packaging, for example buying a refill soap pouch, rather than a new bottle each time. Consumers often think of plastic around food when considering reuse; however, packaging around foods can have benefits and will keep food fresh for longer. So if it is food waste you are looking to reduce, consider whether you will need the packaging to extend the life of the food.

Reuse

Reusing involves taking products or packaging you are considering throwing away and reusing them instead. For example, you could donate old clothes and toys rather than throwing them away or reuse a plastic water bottle.

Recycle

Recycling packaging is covered throughout this book and should take place after reducing and reusing have been considered.

Where do you put bottle lids?

Many people do not know what to do with the lids on glass or plastic bottles. You would be forgiven for putting metal lids in the metal recycling bin, but the trick is to put them straight back on the empty bottle!

Regardless of whether the cap is made of metal, like the ones found on glass bottles, or plastic, like on a soft drinks bottle, the best thing to do when the drink is finished is to pop the lid back on and put it in the correct bin for the bottle material. Yes, this means metal can sometimes go in the glass bin.

Put a lid on it

Small items like lids are generally not very effectively recycled, similar to small pieces of packaging like coffee pods (page 149). A good rule of thumb is that recycling is easier with items larger than a tennis ball. Remember, a lot of sorting at the MRF is based around item size, with small items literally and metaphorically falling through the cracks. This means lids can (and are likely to) get missed by the sorting and recycling process, as they can fall down the various gaps or off the numerous conveyor belts.

Metal lids are normally made from either steel or aluminium (like cans or tins) and glass bottle recyclers will extract them from the crushed glass. Another more time-intensive method of ensuring those caps get recycled is to use a magnet to check whether they are made of aluminium or steel, then fill a can or tin with them and cover it with foil. The main thing is not to put them in loose.

Plastic lids may look the same as the bottle, but they are normally made of a harder type of plastic to ensure the lid is strong enough to stay on. However, plastic recyclers are good at separating the lid from the bottle by shredding it and using density separation (page 71) to split up the different plastics.

There is another advantage to putting lids back on bottles: it reduces the risk of contamination (page 53) in a co-mingled collection (page 46).

What is the scrunch test?

The scrunch test is a really simple way of testing whether a product is made primarily from paper or if it is actually plastic coated in metal, which can look and feel a lot like paper or foil but is unrecyclable. This test is

particularly useful on things like wrapping paper.

It works exactly as it sounds: simply scrunch the material into a ball in your hand, then open your hand – if it has kept its shape, it can be recycled. If it expands back out again, it is not recyclable. The scrunch test works because metal and paper are stronger than plastic film so will hold their shape when scrunched if there is not too much plastic.

Screw it

You will find metallised plastic film on crisp packets, chocolate wrappers, wrapping paper and even helium balloons. The process was invented in the 1930s to make Christmas tinsel, and by the 1970s it had made its way into the packaging we now use every single day. The material is a composite (page 107), which makes it difficult, although not impossible, to recycle.

Aluminium is typically the metal of choice, due to its lightweight property and common use in the packaging industry. The metal is vaporised and effectively sprayed onto cold plastic, turning it back to a solid. Using a metal gives the packaging or wrapping paper a glossy sheen, like foil, but at a reduced weight and cost. Metallised plastic film does more than just look good – it increases the shelf life of a product by preventing air and moisture from permeating the surface.

Foil on its own is infinitely recyclable and does not lose quality. By contrast, metal-coated plastic is not currently widely recycled so it must usually be put in the normal waste bin. However, there are recyclers that will recycle the material collected through specialist collection routes (page 52).

What happens to liquid left in bottles?

When you go to recycle your bottles (drinks, shampoo, soap, anything) the advice in this book is always to empty the liquid out of the bottle and clean it out, recycling only when dry. But what happens if the liquid is still in there? Can the bottle still be recycled?

No rest for the liquid

Plastic bottles put out for recycling are best clean and dry. In theory, they can still be recycled with liquid left in them but there is a risk they will contaminate and therefore affect the recyclability of other materials they are mixed with.

Bottles are normally put out with other materials, sometimes even cardboard and paper, depending on the local authority's collection rules. If liquid is left in the bottle it could stain the card or paper, which could prevent the card or paper from being recycled. Leaving bottle caps on can help prevent this (page 100).

The whole principle of traditional recycling is to take a single stream of materials and process them until they have become something new. During that process, any foreign contaminants will devalue the resulting material, making it harder to sell back into manufacturing.

Liquid inside bottles will also cause the bottle to be heavier. Interestingly, automated sorting machines at recycling plants may use weight to decide what type of material a particular piece of packaging is, so a bottle containing liquid could prevent it being identified as plastic.

Clever deposit machines (page 56) in Norway have been programmed not to give a deposit back if the bottle has liquid left in it. The machine will still take the bottle, but the deposit will go to the shop owner who will have to empty the bottle.[1]

Milk delivery or shop bought?

A question that sums up the complexity of packaging is whether it is better to buy a plastic single-use milk bottle from a shop or to get a delivery of milk in a glass bottle that will be washed out. The answer depends on what you are measuring.

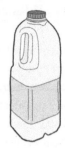

No use crying over spilled milk

Plastic bottles bought from a shop will be made of HDPE (page 122), with a coloured lid made of the same material. This is a widely recycled material and collected from over 75 per cent of local authorities. In fact, a lot of plastic milk bottles now contain around 30 per cent recycled content. The fact that milk bottles are made of natural-coloured HDPE makes them very recyclable.

Doorstep milk deliveries are generally provided in glass bottles, normally with a foil lid. The glass is left out when finished and collected for reuse. These bottles are washed out and reused by the dairy industry. The bottles are reused on average between thirteen and eighteen times before they are recycled.

Let us consider the weight of the packaging to the volume of milk. Glass milk bottles normally contain about one pint of milk, whereas their plastic alternatives can hold up to six pints. To obtain the same quantity of milk, multiple glass bottles would be required. The weight of the glass alone adds significantly to the environmental impact associated with transporting the milk.

A glass bottle will be reused, say, around fifteen times. Research from WRAP (Waste and Resources Action Programme) shows that a glass bottle must be reused twenty times for it to be environmentally equivalent to the plastic.[2]

If we compare the low likelihood of glass being reused twenty times, to the 76 per cent recycling rate of HDPE milk bottles,[3] it feels like plastic is the clear winner!

However, remember, there is no perfect packaging; there are too many variables to be certain which is best. Take, for example, milk deliveries, milk produced and delivered locally compared to complex and long supermarket supply chains. Alas, life-cycle analysis of packaging rarely (if ever) considers all the variables.

It is a great example of complexity though, and it is probably fair to assume that if you usually buy one pint of milk at a time and the milk company reuses the glass bottle on average more than twenty times, glass would be better. If not, you are better off sticking with plastic.

Should you crush cans and squash bottles?

There is something therapeutic about standing over the bin with an empty can and crushing it in your hands – the moment in life we feel most like the Incredible Hulk.

Squashing plastic bottles and crushing aluminium cans is useful as it reduces the amount of space taken up on the collection truck, and you will also be able to fit more in your bin. Unfortunately, it can also pose a problem, as the MRF (page 66) might not be able to tell what the packaging is during the sorting process.

Squash the squash bottle

How can you tell what the rules are where you live? Well, remember MRFs are designed to assess a piece of packaging, which can be done by weight, composition or shape. Changing the shape of a standard product, like a bottle or a can, could affect the ability of the MRF to sort it. There are two main issues that arise from flattening packaging: either the material could be identified as paper, and may accidentally end up in the wrong place, or the air jets that separate waste may struggle to send the recycling to the right area.

With this in mind, the best way to tell whether to squash or not is to find out how your local authority collects its waste. If it mixes everything together ('co-mingled' – page 46), then the waste will be sent to a MRF, so should not be squashed. If you have to sort the waste at home or your council does kerbside sorting (page 79), then you are fine to squash, as this will help fit more in the bin and help the recycler. This rule also applies when recycling 'on the go': if a bin is labelled 'mixed recycling', make sure you do not squash or crush.

When deposit return schemes (page 56) are introduced in the UK and use automatic machines, cans and bottles must be inserted into the machines without being squashed, so that the machine is able to read the product barcode.

What is a composite material?

A composite material is the result of joining two or more materials with different physical or chemical properties. The materials exist in their individual states, rather than being combined, for example when one material is dissolved into another. Concrete is the most common artificial composite material, a combination of stones and cement.

Examples of composite packaging you may come across at home include crisp tubes, drink cartons (page 155) or metallised plastic film.

A connection for life

Why stick two materials together in the first place?

Composite materials are designed to make the most of the best properties of two or more materials and can therefore make packaging stronger and improve the shelf life of the products contained within it. For example, a juice carton may be a combination of paper (75 per cent), plastic (20 per cent) and aluminium foil (5 per cent), to protect the contents from oxygen, light and moisture.

Composite materials that are normally found joined together in packaging include: plastic and metal (metallised plastic film or cans and tins that have a thin plastic liner); cardboard, plastic and aluminium (soup and drink cartons); and paper and aluminium (crisp tubes).

Drink carton manufacturers have invested heavily in the collection of composite cartons. However, as the materials are stuck together, they are generally difficult to recycle and so instead may end up being burned for energy (page 197).

Why can't black plastic be recycled?

Black plastic is used for some fruit, ready meals and meat packaging. Most black plastic is pigmented with something called 'carbon black'. It is seen as a premium colour, which makes bright foods stand out on the  shelf and allows for imperfections in the plastic to be hidden.

Hidden problems

Every year in the UK approximately 3.5 million tonnes of plastic end up in a landfill because it is black; around 1 million tonnes of this is packaging.[4]

Plastic is sorted using near infrared scanning, or NIR (page 71), which detects the type of plastic through its unique wavelength signature. Unfortunately, carbon black pigments are so dark they cannot reflect the infrared beam. This means they are not registered by the NIR detectors and are likely to be missed in the sorting process, ending up in the residual waste and either taken to landfill or incineration. Clear plastic and detectable colours do not have this problem, so there is a call to phase out the use of the pigment.

Carbon black pigment is used in photocopiers and ink cartridges – in fact these very words are likely to have been printed using carbon black. It is made by the incomplete combustion of oil and looks a bit like the soot you might find in a chimney.

The pressure to ensure packaging is recyclable means the use of carbon black in packaging is reducing. Retailers and manufacturers have looked at new ways of packaging products and discovered new types of black pigment that are detectable by NIR, as well as shifting towards using lighter colours. Changing the colour of packaging can cost significantly more, due to the low cost of carbon black, but it is worth it to ensure we do not lose plastic from the recycling process.

What is anaerobic digestion?

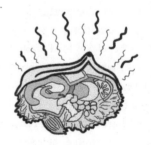

More and more local authorities are collecting food waste, and from 2023 it is likely to be a legal requirement to introduce separate food waste collections. This is important as it makes sure food waste, which is heavy and causes contamination, does not end up in the recycling or waste bin. Anaerobic digestion is the process used to turn the collected food waste into energy.

Food waste bugs me

The term 'anaerobic' refers to a process that takes place in the absence of oxygen. You may have heard of anaerobic exercise, which refers to the breakdown of glucose from intense exercise when oxygen levels are low. Therefore, for anaerobic digestion to occur there must be no oxygen (or light) present during the process.

The first step is to remove any packaging from the food waste, as this will not break down. Biodegradable plastic (page 222) may be designed for this process; however, most recycling plants do not separate compostable and normal plastic when they are removing it and sometimes it is impossible to detect, so it will normally just be removed even if it can undergo anaerobic digestion. This is the main problem for biodegradable plastic, based on current sorting methods. However, as packaging cannot always be removed from food waste, it is better to use biodegradable plastic as a caddy liner as this will degrade in the right conditions.

The 'clean' food waste that is left over is fed into a digester tank, which is then gently heated and stirred. The digestion takes place with the help of bugs and healthy bacteria: they eat the food and create gases rich in methane. This gas is captured and burned, creating renewable energy for homes.

Every tonne of food waste recycled using an anaerobic digestion process instead of landfill prevents around half a tonne of carbon dioxide entering the atmosphere, one of the many benefits of anaerobic digestion.

Pumps or triggers?

Liquid soaps and cleaning products will generally be topped with a pump or a trigger, depending on the product. These clever pieces of plastic transport liquid from the bottom of the bottle to the top, without having to remove the lid.

Pumped to recycle

Triggers are normally found on cleaning sprays and are squeezed like a trigger on a gun; pumps are normally used on bottles of soap and are pushed down.

Spray bottles with triggers were first launched in the late 1960s. A pump draws liquid up a tube, forcing the liquid through a nozzle. Liquid soap and dispensers were used in hospitals throughout the nineteenth century, but it was not until 1980 that they were introduced to the household market.

Triggers are made from plastic, with a complex mechanism inside to spray the liquid. Sometimes these lids are not made to be removed from the bottle, so it used to be difficult to know what to do with them. Luckily, the latest advice on plastic triggers is to put them in the recycling bin with the bottle. This is because the trigger mechanism from most major brands, which used to have metal in it, is now entirely plastic. Keeping the lid on also helps to reduce contamination.

Pumps are also made mostly of plastic but work differently to triggers. They normally contain a metal spring, which causes the pump to lift back after it has been pushed down. This makes them very difficult to recycle. Therefore, if the top requires you to push it down, rather than squeeze a trigger, take the top off the bottle and put the pump top in the normal bin. Then remember to wash the bottle and allow it to dry before recycling it. Alternatively, you can often buy refill bottles that do not include a pump, allowing you to reuse the old pump on a new bottle.

What are microbeads?

Microbeads are small pieces of plastic, usually less than 5 mm in size, although countries do have different definitions relating to the size of a  microbead. They are normally made of PE (page 122) and have been commonly used in toiletries as an exfoliant.

Microbeads, macro problems

Microbeads are washed down the drain and, due to their minute size, can get through sewage treatment plants and into the ocean. Microbeads can look like food to fish; if fish eat the plastic, not only is it detrimental to the health of the animal but the plastic could then also enter the food chain. Microbeads do not degrade over time, but there are natural alternatives to them which will degrade when washed off.

These tiny pieces of plastic were added to shower gels, toothpaste, scrubs and other toiletries. They were designed to help with skin or tooth cleaning. It is estimated the UK cosmetic industry used 680 tonnes of microbeads per year. Unbelievably, a single morning shower could release around 100,000 plastic particles into the sewage system![5]

In 2018, these concerns led the UK government to join a growing list of countries in banning the use of microbeads in toiletries. Their manufacture was prohibited in January 2018 and their retail from July 2018. Interestingly, glitter is by definition a form of microplastic, yet it is not currently banned in craft products, as the UK only focused on cosmetic and personal care products. However, many retailers are removing glitter from things like cards and wrapping paper.

There are lots of alternatives to microbeads that can be used, such as oils, salts, oats and seeds. These natural alternatives will not harm marine life in the way that microbeads may have done.

Can you put recycling in bin bags?

Plastic bin bags are usually made from LDPE (page 122). This type of plastic is flexible plastic film, the same as you would find in carrier bags and bags used to hold fruit and vegetables. As we will discuss numerous times in this book, LDPE is difficult to recycle and so it should not normally be used to hold recycling.

Bin the bag
The main reason to use liners with rubbish is to prevent liquids and food leftovers from getting all over your bin. This is appropriate for your general waste or food caddy, but if you

are recycling correctly and washing and drying out packaging before putting it in the recycling, a bag should not be needed.

Recyclers generally want to sort waste quickly and easily – time is money, after all. If waste is in a black bin liner and hard to see, a lot of local authorities will assume it is general waste and your recycling could end up in a landfill or an incineration site. If a black bag full of recyclate does get to a MRF (page 66), the facility will have to split it open before sorting and remove the bag, as the LDPE is unlikely to be recycled and the carbon black pigment will prevent it getting sorted anyway (page 108).

A lot of bin liners are actually made from recycled waste material. Most retailers collect clear LDPE from their back-of-store waste when they unwrap pallets and boxes. This stream of plastic can be recycled due to its colour and cleanliness, and is sent for recycling into bin liners and other films.

Out and about you may notice that street recycling bins have bags in them. These are usually clear, and they are needed because packaging used 'on the go' is more likely to be contaminated with food and drink as there is nowhere to wash it out.

What do you do with labels?

Labels are found on lots of packaging, such as bottles, jars, boxes and tins. Labels are easy to print on and allow companies to display their brands and product information without needing

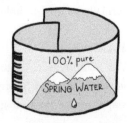

to print directly on the packaging. These do not need to be removed before recycling.

Does what it says on the label

Labels are not just for branding. Much of the information on them is a legal requirement under various pieces of food legislation, and includes things like nutritional information, ingredients and allergens. The fact that the label has to legally reach the consumer can mean they are stuck on with strong glues, potentially making them difficult to remove before recycling.

Fortunately, during the recycling process, when metals and plastics are heated, the labels and the glue used to stick them down will burn away and be used for energy recovery. When glass is crushed, non-glass items like labels are removed using air and the washing process. When the glass is melted, the heat will also remove any glue. With cardboard boxes, the labels will be pulped with the fibres during the recycling process and any contaminants like ink will be removed.

Sometimes drinks companies use a plastic sleeve which covers the whole bottle. In theory these can be recycled, but a sleeve which covers more than 60 per cent of the bottle will cause a problem with the sorting process. Near infrared scanners which identify plastic by its material type can get confused by a sleeve covering the bottle and read the properties of the plastic sleeve rather than the bottle (page 71). As these can be made of different types of plastic, the bottle could be misidentified, impacting the recycling process. With any bottle that is largely covered by a sleeve, it is important to remove it before recycling.

What does 'circular economy' mean?

'Circular economy' is a term used more and more frequently to describe the aims of recycling. A circular economy will keep materials in use for as long as possible. A good example of a 'circular economy' is an electrical firm that deliberately designs its products to be easily taken apart and manufactured with components that can be replaced, thus promoting the continuation of a potentially single-life product. Another example is a reusable coffee cup being used and washed out for your daily caffeine intake.

Around and around

The natural world is cyclical – consider a water droplet evaporating and turning back into rain. The 'circular economy' concept aims to do this with products like mobile phones or plastic, by designing them for reuse or easy collection and recycling.

In 1966, an economist named Kenneth Boulding introduced the idea of an 'open economy', where resources are lost, and a 'closed economy', where resources exist for as long as they can. Since the 1980s this concept has evolved into a key component of environmental frameworks around the world.

A circular economy is not just focused on recycling. It will also focus on sharing, repairing and reusing; anything that prevents waste from being created. By buying things and not putting them back in the loop at the end of their life, we use up resources that will not be around for ever. Resource extraction can create toxic waste and is not sustainable in a growing economy.

Until relatively recently the world has largely existed in a 'linear economy', where products and packaging are made, used and disposed of. As awareness of the environmental damage of this type of consumption grows, it is essential that our global approach becomes more circular and focuses on recovering materials back from the people who buy them, so that they can be used again.

Your recycling knowledge will have increased significantly by now, but there are particular recycling symbols that we haven't explored yet. Some are more useful than others, as we will find out in the next section.

7

The Rubbish Symbols

What is the Green Dot?

The Green Dot is a symbol with two intertwined arrows. It can be found on most packaging, particularly products that are sold in Europe. However, unfortunately it has no practical meaning in the UK.

Connect the dots

The Green Dot was created in Germany in 1990 to signify to consumers that the company that made the product had contributed towards the cost of its recycling.

Its use is managed by an organisation called Packaging Recovery Organisation Europe (PRO EUROPE). It is licensed by thirty-one countries in Europe,[1] although in the UK and the Netherlands it does not demonstrate that a financial contribution has been made. Historically, the Green Dot was mandatory in lots of countries, but this is changing, and it is now a more voluntary (albeit still widely used) scheme. The Green Dot remains mandatory in Spain and Cyprus.

Current legislation in the UK includes producer responsibility (page 14), a concept introduced by a European directive putting an obligation on producers to recover, sort and recycle packaging. Each country in Europe implemented the directive differently, and the Green Dot indicates that a producer has contributed towards their respective version of the legislation and that the cost of recycling the packaging has been provided for.

The Green Dot remains prevalent in the UK because many companies want to avoid printing two separate sets of packaging for different countries, which would not be very environmentally friendly!

As a trademarked logo, companies in the UK do pay a small amount of money to use it, but its presence does not signify anything in terms of recycling.

What is the Mobius Loop?

Perhaps the most recognisable of all the Rubbish Symbols, the Mobius Loop features three clockwise arrows chasing each other in a triangle shape. You will find it on a lot of packaging; it is often recognised, although its meaning is not widely understood.

Infinite confusion

Like 'reduce, reuse, recycle' (page 98), the Mobius Loop has its origins in the first Earth Day on 22 April 1970. Gary Anderson, a twenty-three-year-old student in North Carolina, won a sponsored contest to design a symbol that would raise awareness of environmental issues with his Mobius Loop. Today, the symbol is not trademarked and is free to use in the public domain, hence its appearance on packaging around the world.

The Mobius Loop indicates that the packaging is made from a material that can be recycled. The confusion surrounding the Mobius Loop comes from the fact that people assume it means a product can *always* be recycled. However, local authorities do not always collect the same packaging (page 49), so it is not as clear-cut as that. In this regard, the OPRL label which assesses UK collection coverage is superior (page 133). After all, virtually any packaging *could* be recycled if you could collect it on its own and were not worried about the economic viability of scaling the recycling process.

There is a difference between packaging that is made from recycled material, and something that can be recycled. Sometimes, the Mobius Loop will have a percentage or a number in the centre of it. The percentage signifies the proportion of material in the packaging that has come from recycled content. A number is used on plastic to signify the plastic resin from which the packaging is made, as we will explore next.

What do the seven plastic numbers mean?

Pick up any piece of packaging made of plastic, look really closely and chances are that somewhere you will find a triangle imprinted into the surface containing a number within it. Never noticed that before? You are certainly not alone.

The numbers range from 1 to 7 and indicate the type of plastic you are holding. They are useful when buying products as different numbers have differing levels of recyclability; some of them are definitely more economical to recycle than others.

The system of identifying plastics by numbers was born in America in 1988. Known as the 'resin identification code', it was designed to promote the consistent recycling of post-consumer plastic. The sorting of plastics was key to their recycling, and this code made the job easier.

The names originate from the chemicals involved in the production of plastic. Fortunately, they are known by their initials, which are far easier to remember!

Number 1 – PET (Polyethylene Terephthalate)

Outside of packaging, you will have come across PET by its more common name: polyester. PET is generally used to refer to packaging and it is one of the most commonly used plastics.

This is the type of plastic used to make drink bottles and it is highly recyclable, hence the desire to ensure it is collected through deposit return schemes (page 56).

Nearly all PET recycling is mechanical (page 88) as chemical recycling, while possible, is not viable at scale with the relatively easy-to-recycle plastics.

Number 2 – HDPE (High-Density Polyethylene)

Part of the polyethylene (PE) family, the 'high-density' of HDPE means it is a very strong plastic. It is used to make bottles for milk, shampoo and detergent.

Its specific use, particularly as a milk bottle, ensures it is recycled widely. The key to successful and economical recycling is to reduce the variants of packaging, and since most of the milk industry uses HDPE, this leads to a very high recycling rate of this plastic stream. HDPE's focused use also means the plastic can be recycled back to food-grade plastic, resulting in a closed loop for the dairy industry.

Number 3 – PVC (Polyvinyl Chloride)

PVC comes in two types, rigid and flexible, and is made from a mixture of oil and salt.

As a soft, flexible plastic, it is used to make products such as cling film. Rigid PVC is used to make products including doors, windows and pipes. In fact, vinyl records are so named due to their PVC composition.

In its flexible form (which is more widely used in packaging), PVC is very difficult to recycle because of its high level of chlorine. It is therefore very unlikely to be collected and recycled in a mixed plastic stream. However, non-PVC flexible films are available (page 159).

Rigid PVC, which is cheap to make and weather resistant, is more likely to be recycled as it will usually be removed from a house by a business (in the form of windows or piping), allowing it to be sent for recycling. PVC can be very dangerous if it is burned as it can release harmful chemicals, so in its rigid form it is important that it is recycled appropriately at the end of its life.

Number 4 – LDPE (Low-Density Polyethylene)

Like HDPE, LDPE belongs to the polyethylene (PE) group. This is the flexible version of polyethylene plastic, hence the 'low-density' in the name.

LDPE is mostly used to protect food. You'll find it wrapped around fruit and vegetables or as the key component of bread bags. A 'bag for life' would also be made of LDPE.

Due to the lightweight nature of the material, it is not usually cost effective to collect it from the kerbside and transport it to the recycler. However, fortunately, most large supermarkets now have carrier bag bins where you can drop off any LDPE for recycling. It will typically go on to become a new carrier bag or film. Number 4… take it to store!

Number 5 – PP (Polypropylene)

PP is tough, lightweight and does not mind heat. It keeps products dry and fresh, so it is used for things like the bag inside your cereal box. This also makes it an effective packaging for margarine tubs, plastic straws (before they were banned) and yoghurt containers.

Around 70 per cent of PP is estimated to be used in food contact applications and a lot of work is currently underway to increase the recycling of PP back into food-grade plastic. The technology exists, but this does not currently take place at scale.

As a packaging type, PP is increasingly collected from the home but there can be significant differences between local authorities, so you should check your council website.

Number 6 – PS (Polystyrene)

Like PVC and the types of polyethylene, polystyrene comes in two types: 'expanded', or a solid, clear plastic.

'Expanded polystyrene' is the plastic that gets everywhere when you buy a new TV and is used a lot in containers from takeaway restaurants. It is the white or beige coloured packaging that breaks easily into pieces and makes a real mess if it does. The challenge with recycling the material is that 95 per cent of it is made of air and so its bulk and low weight makes it inefficient to move around. However, when compacted it can be commercially viable to recycle it. This means that, interestingly, it actually has a very high recycling rate between businesses, but not from households.

As a rigid plastic, polystyrene might be used in things like yoghurt pots, although most companies are now moving away from using it and towards PP, due to polystyrene's low collection and recycling rate.

Number 7 – Other

Any plastic material that does not fit within the other six categories falls into number 7, which includes plastics such as acrylic, polycarbonate and bioplastics.

Historically, you would not find number 7 in normal day-to-day plastic packaging as it is usually used to make products such as iPods, sunglasses, computers and some reusable water bottles.

However, newer bioplastics (page 221) are included in this category. Putting a whole plastic stream in a catch-all category can create quite a lot of confusion during the sorting and recycling process. To help, sometimes these plastics will have the letters 'PLA' (which stands for polylactic acid) near the number to identify this specific stream.

What are the symbols for glass, aluminium, and steel?

The Rubbish Symbols can be used to distinguish different types of material. This is particularly useful with metals, as aluminium and steel look similar.

The symbols below can be found on packaging. However, their use is not mandatory and so they are usually found less frequently than the plastic symbols (page 122).

Glass

This symbol is used as a reminder that you should recycle glass packaging.

The image resembles a person putting a bottle in a bottle bank (page 58), although today glass will normally be collected and recycled from kerbside collections.

Aluminium

This symbol shows that the packaging is made from recyclable aluminium. In a recent study, this symbol was found to be the most confusing to the consumer.[2] You will find it most commonly on cans.

Steel

This symbol means the product is made from steel. The symbol is a tin can being attracted to a magnet, reflecting the key sorting property of steel (page 70). This symbol will mostly be found on tin cans.

What is the Tidyman symbol?

The Tidyman is a very familiar symbol found on packaging and the figurehead for anti-litter advertising. A stylised figure leaning over a bin to drop some rubbish in, it was launched in the late 1960s, and over time has become a nationally recognised symbol reminding everyone not to litter.

Litter logo

The symbol was launched in the UK by the charity Keep Britain Tidy. Its official origins remain unclear; many cite a partnership between Keep America Beautiful and the United States Brewers'

Association, although neither have records to show that they were the original creators of this sixty-year-old icon.

Research conducted by Keep Britain Tidy suggests that more than eight out of ten adults recognise the symbol and understand what it means, making it one of the best-known symbols in the UK.[3] Tidyman did spend a bit of time out of the public eye but has been relaunched in anti-litter campaigns since 2017.

The Tidyman logo does not actually relate to recycling or whether an item can be recycled. It is included on packaging and bins purely as a reminder to place items in the bin when they are finished with.

Despite Tidyman being a well-recognised symbol, littering remains a significant problem in the UK. Therefore, while symbols like this which prompt the public have been important in the past, now that consumers are more engaged than ever with recycling and the environment it is important that symbols do more than just prompt and actually help us determine whether something is recyclable and how it is likely to be recycled.

What is the symbol for electricals and batteries?

Unsurprisingly, this is called the 'crossed-out wheelie bin' symbol. It is a legal requirement for an importer or manufacturer of electrical goods or batteries to print the symbol somewhere on the product. It serves as a reminder not to throw electricals

or batteries into the bin and to recycle them properly instead.

Crossed wires

Introduced in 2005 as part of the creation of the European directive on waste from electrical and electronic equipment (page 14), it has since been a legal requirement for the crossed-out wheelie bin symbol to appear on electrical items themselves. When batteries legislation arrived in 2009, it adopted the same symbol. You should be able to look at any electrical item or battery and find it somewhere.

The symbol is a reminder that electricals should be taken to a local recycling centre or back to a store (page 60). Batteries should be taken back to any store selling batteries, which are legally required to collect them (page 61). Both waste electricals and batteries contain valuable metals and are hazardous if put in landfill.

In modern electronics, you should be able to find a solid black line under the crossed-out wheelie bin. This is used to indicate when the electrical item was sold. It is a legal requirement for any electrical item made after 2005 to have the black line underneath, as this is when the legislation was created. Unfortunately, if you look at some modern (and common) electrical items, including those from leading mobile phone companies, you may not see the black line under the logo. Even though it is a mandatory element of the logo, apparently it is less aesthetically pleasing, but the policing of the symbol appears to be non-existent as companies still fail to meet the requirements.

What is the symbol for compostable packaging?

The symbol for compostable packaging is a plant growing out of a loop of plastic. This symbol, known as 'seedling', is a registered trademark of European Bioplastics and indicates that the packaging is certified as compostable (packaging that will break down and return nutrients back to the earth).

Break it down

In order to decompose, compostable packaging may be made of a bioplastic and therefore may be stamped with the resin identification code 7 (page 122). However, the term 'bioplastics' simply refers to the chemical make-up of the plastic, and as such not all bioplastics are compostable, hence the need for a separate, specific symbol.

This book covers compostable packaging in more detail later (page 225) including its advantages and disadvantages as a packaging type. For now, it is important to understand that one of the major issues with compostable packaging is signifying that it is compostable, so that it is put in the correct bin. Otherwise, there is no way of determining whether a compostable plastic, which looks and feels like oil-based plastic, is indeed compostable.

For this reason, the EU has created a threshold (snappily referred to as EN 13432/14955) which compostable packaging must meet in order to use this logo. This standard sets out how quickly the packaging should break down under certain conditions.

The only downside to this symbol and standard is the word 'compostable'; by definition, it might make consumers think they can compost the material at home, when in reality most compostable plastic will only break down in controlled industrial composting facilities.

What is the symbol for home compostable packaging?

The compostable packaging symbol we have just looked at is intended for packaging which will be composted industrially, and so should

be put out with food waste for collection. There is a different symbol for packaging which can be composted at home, rather than taken away.

Homeward ground

Compostable packaging will break down under controlled conditions (including a specific temperature, level of humidity and aeration), so it is important to know whether such conditions can be achieved in a compost bin at home, rather than an industrial composter. This symbol was created to indicate when a product is suitable for home composting.

In theory, home composting is easy to do, and environmentally one of the best ways of getting rid of organic waste. Your home compost bin, which will sit in the garden, can take food peelings and scraps, grass cuttings, eggshells, pet waste and ground coffee or loose tea. However, with the different types of compostable packaging, it is a confusing area.

This symbol, created by a certification body called TUV Austria over seventeen years ago, is known as the 'OK compost home' symbol. It is a certification of technical requirements that packaging must meet in order to be home compostable. Essentially, it means the packaging will break down at a lower temperature and over a longer period of time.

The variability of a home compost bin, in temperature for example, makes home composting much harder to standardise than industrial composting. For this reason, there are no specific standards for home compostable packaging, and it should be noted again that most compostable packaging in the UK is designed for industrial composting, not home composting.

What is the OPRL symbol?

OPRL is a not-for-profit organisation whose label has become a very common way for brands to communicate the recyclability of their product. OPRL stands for 'on-pack recycling label'. Most major brands and retailers have  signed up to be part of the scheme so you will find this logo on lots of everyday packaging in the UK.

Historically, the symbol was designed to tell you whether packaging is likely to be collected for recycling, whether it should be taken to a local recycling centre or thrown away. In 2020, the scheme was simplified significantly to 'recycle' or 'don't recycle', with some additional specialist labels. It will take time for the new system to flow down to the products you buy so you might notice the older 'widely recycled', 'check local recycling' or 'not currently recycled' symbols for the time being.

Recycle

This label is found on types of packaging that more than 75 per cent of councils will collect from home; the material will then be sorted, recycled and must have economic value.

Specialist label

This label is found on packaging where more than 75 per cent of the population could have easy access to a collection facility (not necessarily kerbside) and the material will be sorted, recycled and hold economic value.

Specialist labelling might be found on a coffee cup, which would be labelled 'recycle at coffee shop', or LDPE (page 122), whose label would say 'recycle with bags at a supermarket'. See specialist collections (page 52).

Don't recycle

Everything else gets a 'don't recycle' label.

Refil-label

In 2021, OPRL started broadening its scope, moving into new areas which surround, but are not directly linked to, recycling.

The first label to be released in this broader remit was the 'refill' labels, designed to help consumers understand when packaging can be reused. Three labels have been released: refill at home, refill in store and return to refill.

To qualify for the label the packaging must have been designed for reuse a minimum of ten times, with refill systems available to 75 per cent of the UK population for a minimum of three years.

This evolving labelling, giving the consumer more information, marks an improvement to the current system and helps to move packaging higher up the waste hierarchy (page 96).

What is the FSC symbol?

FSC stands for 'Forest Stewardship Council', an organisation which manages FSC certification. This symbol is used for wood, paper and cardboard and indicates that the wood sourced to make the product comes from well-managed forests, meeting the highest environmental and social standards.

The best things in life are tree

The Forest Stewardship Council was established in 1993 in Toronto, Canada by a group of businesses, environmentalists and community leaders. The FSC now operates in over eighty countries and its headquarters are in Bonn, Germany.[4]

There are two types of certification: forest management, which certifies the forest at the start of the supply chain; and the chain of custody, which passes the certification along the supply chain from the forest to the final product.

In order to be given FSC certification, a forest must be managed in a way that looks after the environment. There are three types of label: FSC 100%, FSC Mix or FSC Recycled.

FSC 100%

The wood comes completely from FSC-certified forests.

FSC Mix

Established in 2004, this certification allows wood from FSC-certified forests or recycled material to be mixed with non-certified wood in controlled conditions.

FSC Recycled

The wood or paper comes from 100 per cent recycled material (whether or not the original material was FSC-certified).

That completes all the major symbols you may find on packaging and products you buy. The next section will stop you hovering over a bin wondering if something is recyclable. It's time to get an A-Z of products.

8

The Rubbish Encyclopedia

We have taken time building up our knowledge on the world of waste; however, there is still a chance individual products and packaging may cause confusion. This section is a list of 50 products, in alphabetical order, to help.

Batteries

As we have covered, batteries are collected (page 61), usually from shops, and then sent for recycling (page 92). All batteries can be recycled but certain types are harder to recycle than others.

No charge

Any shop that sells a pack of four batteries a day must have a battery recycling point and employees must be aware of it. Next time you go into a shop, have a look and see if you can find the battery bin. If not, you can ask them to get one because, by law, the shop is required to collect batteries for free. This legal requirement has led to 45 per cent of batteries sold in the UK being recycled.

As a consumer, your household batteries will be recycled for free; the recycling is funded by the companies that manufacture batteries. The appropriate collection and recycling of batteries is really important as they are susceptible to catching fire if they are crushed in a normal kerbside collection truck (page 199).

Some councils do collect batteries, so it is worth checking whether yours does. They will normally require you to place them in a clear bag. You could use a bread bag for this, rather than buying something new.

Most household batteries are exported for recycling. UK-based recycling infrastructure is likely to develop due to the increasing costs of exporting recycling, and the environmental impact it entails.

Birthday cards

Most greetings cards can be recycled, as they are no different from paperboard. Watch out for those decorations, though; glitter, ribbons, anything shiny... generally a nightmare to a recycler!

Glitter = litter

Greetings cards were first given in Europe during the fifteenth century, although the Egyptians and Chinese had been exchanging notes of well wishes long before that. The 1850s saw the transformation of cards from something highly personal and expensive to an affordable communication method. It is estimated that in the UK alone nearly 1 billion cards are sent each year.[1]

The addition of glitter and non-paper decorations is a fairly modern problem. Any decorations on a card that are not paper based are likely to contaminate recycling, so it is important to remove them where possible.

If the card is covered in glitter which cannot be removed, it is best to rip off the section with glitter on and just recycle the remaining card. Better yet, ask people not to give you cards with glitter, which are essentially tiny pieces of plastic which will never get recycled due to their size. For this reason a lot of retailers are now moving away from adding decorations such as glitter to their cards.

Been given a card that makes a sound when you open it? Make sure you remove the battery and electrical parts before it is recycled. The battery can be taken to a battery recycling point, which you can find in most shops (page 61).

Biscuit and sweet tins

Biscuit and sweet tins are containers, usually made of steel and full of treats. Mainly bought and used around Christmas, the steel tins are very easy to recycle, should you not wish to reuse them.

Sweet things tinside

The first biscuit tins were released in 1832. These early tins were handmade and had a glass lid. By the end of the decade, tins made in various shapes were commonplace. Today, they are made in the same way as a tin can, using tin-plated steel.

Biscuit and sweet tins are very effective at keeping moisture out as the lid seals tight to the base, which helps keep the product fresh. This also means that when they are finished with they are great for storing things: jewellery, a sewing kit, cake or even more biscuits!

If reuse is not possible, both the tin and the lid can be recycled. Biscuits will often come with black plastic liners which cannot be recycled (page 108). Remember to remove this first and put it in the normal waste bin.

Sweets and chocolate will normally be individually wrapped in plastic. Again, this is unlikely to be recyclable as the wrapper will probably be made of metallised plastic to keep the product fresh. Some manufacturers now use compostable plastic to wrap individual chocolates; although food waste collectors do not want plastic in the food waste, with the exception of the caddy liner, so this means compostable wrappers should go in the normal waste bin.

Bottles (glass and plastic)

Most drink bottles are made of glass or PET plastic. Both are very recyclable and can be used to make new bottles. The introduction of a deposit return scheme focused on bottles (page 56) could create a significant improvement in recycling rates and a true circular economy.

You can leave your cap on

Glass bottles are recycled by smashing them up and removing the lids and any labels. The smashed pieces, known as cullet, are then sorted by colour, and melted down to be formed into new glass bottles. This process can be repeated for ever without the bottles decreasing in quality, which is one of the main advantages of glass recycling.

Plastic bottles are also recycled by melting. The plastic is sorted by type, washed to remove contaminants and shredded into flakes. Finally, the flakes are melted into small spheres called pellets, about the size of a grain of rice. The pellets are then melted before being turned into new bottles.

Unlike glass, the quality of recycled plastic will decrease over time, so bottles are made from a mix of recycled and new plastic. This keeps the quality high, which is important for plastic that is used to contain food or drink. To ensure your plastic bottle has its longest life, the best thing to do is refill and reuse it.

When recycling the bottle always wash it out, let the bottle dry and put the cap back on to ensure quality material is sent to the recycler.

Cans and tins

Cans and tins form one of the most common types of packaging for food and drink. They help preserve products and are made of aluminium (cans) or tin-plated steel (tins), which helpfully means it is possible to recycle them over and over again.

Over 4.5 billion aluminium cans are recycled each year and recycling a single can saves enough energy to power a TV for three hours.[2]

Can do attitude

Cans are usually made from aluminium, and tins usually from steel. Steel is magnetic and aluminium is not. If you own a magnet, you can play the role of a sorting facility by lining up tins and cans and passing the magnet near them. The tins will start getting pulled to the magnet, just as they do in a MRF (page 70).

Cans and tins are recycled using very high heat (page 86) so any liquid and food left in them is not too much of a problem during the recycling process, as they will be burned away. However, to avoid contamination with other materials in the bin such as paper and cardboard, it is essential to give them a wash and ensure they are dry before putting them in the recycling.

Tins are opened with a tin opener or a ring pull. The best thing to do with the lid is put it back in the empty tin once it has been washed out, as this will reduce the chance of it getting lost in the sorting process.

Cardboard

Cardboard is highly recyclable as long as it has low levels of contamination, which unfortunately can include water, so be careful on rainy bin-collection days! As a packaging type, its use is increasing as more shopping is sent to the home in boxes through online delivery companies.

Eggs-cellent boxes

If cardboard is sent to rot in a landfill it will emit methane; if it is incinerated it releases carbon dioxide. Both gases contribute to climate change, so the fact that cardboard currently has a high recycling rate helps to reduce its impact as a packaging material.

Many things are made of cardboard, including cereal boxes, online delivery boxes (usually corrugated cardboard), greetings cards and egg cartons.

Most cardboard can be recycled in a process that involves mixing the cardboard with water and chemicals and then spinning it to create a pulp. The pulp is sprayed in a thin layer before being pressed and dried to form recycled cardboard (page 84).

The only types of cardboard you should avoid recycling are items heavily contaminated with food, oil, dirt or paint. This could stain the recycled product as it is unlikely to be removed in the recycling process. Water can also make it very difficult to sort cardboard as it changes its weight and composition, so make sure to use lids on your cardboard waste bin, if provided.

Carrier bags

Carrier bags are usually made of HDPE (single use) or LDPE (bags for life) and technically can be recycled, but as they are made of lightweight plastic film, collection is difficult. LDPE is the same type of plastic as the film used to wrap goods like fruit and vegetables.

Often, when materials have a low recycling rate, it is actually the collection of the material that is the problem rather than the recycling process. For carrier bag recycling to be successful, it requires special collection points which allow for the bags to be collected in bulk (page 52).

Bring bags back

In order to recycle your carrier bags at the end of their life you will need to take them to a large supermarket for recycling. Most large supermarkets will have a special bin for carrier bags near the entrance of the store.

Since the carrier bag charge was introduced in 2015 (page 203), most people now reuse their plastic carrier bags.[3] The charge was introduced to stop people from taking a new bag or bags every time they shop and has worked very well to reduce the number of new bags being used, although some of this use may have shifted to alternatives like bags for life, which can also be problematic for the environment.

The best thing consumers can do with carrier bags is reuse them. All bags have an environmental impact, and the more they are reused the more that environmental impact is mitigated. As retailers move to thicker bags for life, it becomes even more important that they are reused.

Chemical bottles

You would be forgiven for thinking bottles full of corroding and dangerous chemicals found lurking in under-sink cupboards and bathrooms cannot be recycled. However, they are made of widely recycled HDPE (page 122) so they can be... mostly.

Cleaning up

Bottles containing products like bleach, antibacterial spray and toilet cleaner are found in bathroom and kitchen cupboards. Bottles containing chemicals will be made of HDPE, plastic number 2. As a plastic made with high-density polyethylene, it is corrosion-resistant, making it a practical storage solution for chemicals.

Having said that, if a bottle contains very strong acids, there could be a risk to recyclers further down the recycling chain. A good rule of thumb is if the product can be safely put down the sink (it will state this on the label) it is fine to recycle the bottle.

The bottle will need a good wash before it goes in the plastic bin, so make sure you give it a thorough rinse out and ensure it is dry. Then place the cap back on, ensuring the bottle is sealed, and put it in the plastics bin. As always, leave the label on as this will be removed at the plastic recycling facility (page 144).

Bottles containing stronger chemicals or acids, like antifreeze, which will be labelled to state they must not be washed down the sink, should be put in the normal waste bin once the chemicals have been safely dealt with.

Clothes and textiles

Clothing and textiles include a huge range of products, such as curtains, bedding, cushion and sofa covers. Over 300,000 tonnes of unwanted clothing gets thrown away each year, when in fact this can all be reused in some capacity and should never be thrown away.

A complete fabrication

Fabric is always useful to someone. This is true of old clothing that's been replaced by new styles, and also of textiles that are past their best – perhaps they're sporting a few holes, or some stubborn food stains.

The first step is to check your local authority website, as some of them will collect clothing and textiles in their recycling collections.

Any fabric in good condition can be donated or sold. Lots of charity shops will take textiles; alternatively, you may find textile collection points in large supermarket car parks around the country, near bottle banks or other recycling points.

Some charities actually offer a collection service, so it is worth having a look online to see if your chosen charity does, although dropping it off at a charity shop when you next visit could be a more environmentally friendly alternative and save the charity money.

Some retailers run 'bring back schemes' where you can drop off unwanted clothes for recycling; again, it is worth checking company websites.

All clothing can stay out of the bin. Any clothes that are deemed not suitable for reuse will be recycled and could get turned into fabric-based items like cloths, chair covers or blankets.

Clothes hangers

The average person has a wardrobe full of them, and they have not changed much since 1890 when wire

hangers were first patented. It is easy to end up with more than you need when you have a clear-out, but what do you do with them?

Hang in there

Clothes hangers are made of metal wire, wood or plastic. Unfortunately, all are difficult to recycle and unlikely to go in your household collection, although it is always worth checking.

Wire hangers are normally made of steel. In theory, this could go in the metal bin at your home; however, they are difficult to collect due to their low weight and size. They can be taken to your local household waste recycling centre or they make very useful materials for DIY or craft projects.

Some plastic hangers are black plastic, making them hard to recycle (page 108). They are also likely to be made of polystyrene, which makes them doubly difficult to recycle. This means all broken plastic hangers should go in the normal bin.

If wooden hangers are broken, they will need to go to the wood bin in your household recycling centre.

Intact hangers are perfect for donating to charity shops, so check if your local shop needs some. The best thing to do, though, is to just refuse them in the shop when you buy clothes, leaving them behind to be reused or recycled.

Coffee cups

Disposable coffee cups have been making headlines since 2015, when it was revealed that over 2.5 billion coffee cups are thrown away and not recycled each year.[4]

Recycling coffee cups is harder than their cardboard look and feel would have you believe. Most coffee cups are a composite material (page 107), and therefore have to be taken to special collection points.

A bean-ormous problem

Coffee cups are made of cardboard with a thin plastic lining, most of which is on the inside of the cup to make it leakproof and heatproof, key properties for a coffee cup.

This composite material makes cups difficult to recycle as the plastic has to be separated before the paper can be recycled. Due to increased pressure from the public, the industry has been working very hard and has created multiple collection points where coffee cups can now be collected. In fact, most coffee cups can now be returned to any coffee shop, regardless of brand. Takeback of coffee cups in coffee shops may even become mandatory from 2024, following the introduction of extended producer responsibility (page 216). After collection, cups are taken to specially designed recycling plants where the plastic is separated from the paper before recycling.

Reducing is always better than recycling (page 96), so the best thing to do is switch to a reusable cup, and most coffee shops will now offer a discount if you bring your own cup.

Coffee pods

Around 30 per cent of households in the UK have a coffee machine that accepts pods made of aluminium or plastic, a figure which is growing each year.[5] These pods deliver a single serving of coffee, tea or milk. They are difficult to collect and recycle due to their size and the fact that they retain liquid. However, thanks to investment from the industry, the collection and recycling rate is increasing each year.

Oh my pod!

You might imagine that single-serve pods must be bad for the environment – after all, they are single use. However, much of their environmental impact comes from the coffee itself, not the packaging. Pods are one of the most efficient ways of delivering a cup of coffee as they require significantly less coffee than other methods. This actually makes pods better for the environment than other coffee brewing methods.

The issue with recycling pods is their small size, as they can get lost in a MRF (page 66), and the residual coffee they contain causes contamination during the recycling process.

The industry has recognised the need to collect and recycle the packaging and has in the past developed independent schemes. In 2020, a major scheme was announced called Podback, which unified the way major pod brands collect and recycle the product. The scheme will increase kerbside collection, funded by the industry, and ensure that many more people can easily recycle their pods in the future.

Cotton wool

All used cotton wool is unrecyclable and should be placed straight in the normal rubbish bin.

Cotton is an unusual material, as it gets stronger when it is wet. This is important for the way it is used, but this unique property can lead to it damaging the environment.

Wool-d you bin it?

A plant-based, silky fibre that comes directly from the cotton plant, cotton is used in lots of industries including for clothing, paper and even banknotes, although newer money is now made of plastic. Growing cotton takes up 2.5 per cent of global farmland![6]

Cotton wool is made up of short fibres, similar to those used to make paper. However, when it is used, cotton wool gets contaminated with nail polish remover, make-up or anything else it comes into contact with. This makes it impossible to recycle.

To help keep a specific shape, some cotton wool pads contain plastic, adding to the recycling complexity. There are ethical cotton wool products, but these still cannot be recycled, so it is much better to use something washable and reusable.

As cotton wool is mostly used in the bathroom, it is tempting to flush it down the toilet. However, as cotton wool expands in water, this could lead to it getting stuck down a pipe and causing blockages, so just bin it instead.

Crisp packets and chocolate wrappers

Crisp packets and chocolate wrappers are made of the same material: plastic which has been coated in a very thin layer of metal (metallised plastic film). This material is not widely collected  or recycled, although this could change over time as chemical recycling becomes more widespread (page 219).

Salt and bin-egar

Crisp packets and chocolate wrappers are both made of composite materials (page 107). They also fail the scrunch test (page 101), which means there is not enough metal content in them to suggest they should be recycled in the metal recycling bin. Try scrunching them into a ball and you will see they start to return to their original shape.

Crisp packets and chocolate wrappers are coated in metal to keep the food fresh and to stop air and moisture penetrating the packaging. This layer helps increase their shelf life, stopping the contents from becoming food waste.

As plastic that cannot be recycled has become a significant environmental problem, brands have been exploring new types of packaging that could be more recyclable, avoiding the layer of metal and focusing on more recyclable plastic. There are also future opportunities with chemical recycling or pyrolysis which could break down the plastic and metal, separating the individual components so they can be used in a recycling process.

Crockery and cutlery

This category includes plates, bowls, cutlery and mugs. If they break or are changed for a new set, what do you do with the old ones?

The good news is, if it is hard, such as ceramic crockery or metal cutlery, it can be recycled at your local recycling centre. Plastic crockery and cutlery, however, is usually made of polystyrene (page 122) and therefore is not currently collected or recycled. Disposable cutlery and crockery is unlikely to be recycled, even if it is made of wood or paper, due to its size or high levels of food contamination. This is in the face of marketing typically associated with these items. Many companies have moved away from plastic to alternative unrecyclable products. Sometimes food outlets will have special compostable bins for these; if not, then they are no better than the plastic alternative and will need to go in the normal bin.

Unbeliva-bowl!

Crockery and cutlery should always be reused or given away first. If they are not chipped or broken, they could be donated to a charity shop, given to family or friends or sold. Remember, reuse is always better than recycling, as the manufacture of new items is energy-intensive and this is avoided if a product's life is extended by reuse.

If crockery or cutlery is broken and therefore not suitable for reuse, it is normally accepted at the local recycling centre (although it is always worth checking first). Broken crockery goes in the 'rubble' skip. All metal which is not collected at kerbside, including cutlery, goes in the 'metals' skip.

After it's been put in the skip, any contamination will be removed from the crockery and it will then be crushed for use in gravel or under driveways. Cutlery will be melted down and recycled.

Deodorant and antiperspirant cans

As well as deodorant and antiperspirant, aerosol cans contain all sorts of products, such as hair spray, paint and air freshener.

Similar in composition to cans or tins, they are made of steel or aluminium and therefore are highly recyclable; however, the pressurised air inside makes recycling them a bit more challenging in practice.

Under pressure

The problem with aerosol cans is the pressure that they work under and the gases contained inside which make them spray. This makes the cans dangerous if a hole is punched through them. Cans are pierced safely during the recycling process, and the gas removed. However, if they are thrown in the normal bin the cans can explode when they are crushed, either in the truck or during the disposal process. This is why it is so important to recycle them.

It is very simple to recycle aerosol cans, but just make sure all the product inside is used up first. The best way to do this is to shake the can and listen for any liquid inside, then spray until nothing is coming out of the can.

Do not attempt to put a hole in the can or remove the spray nozzle to get at any product left in the can.

Once the can is finished, remove any non-metal parts that come off easily, like the lid, and put the can in the metal recycling bin. Any plastics still on it that cannot be removed without force will be extracted during the recycling process.

Drink cartons

Drink cartons are usually composite packaging (page 107) made of three types of material stuck together: paper, plastic and aluminium. This makes them difficult to recycle, but thanks to investment from the companies that make the cartons, they are now widely collected, and they have a recycling rate reported to be around 50 per cent,[7] although this has been contested by industry experts who believe it could be as low as 29.5 per cent in the UK.[8] Currently, they go straight in the cardboard bin when finished, if your local authority collects them. This could change, and it may be that they're put with bottles and cans in the future.

Mix it up

First used in the 1900s to contain drink, specifically milk, little thought was given to the environmental credentials of cartons in the early days. However, pressure has led to the industry ensuring 68 per cent of the UK has access to collections.[9]

Cartons are usually collected from the home and then either sent straight to a specialist recycler or sorted at a MRF (page 66), getting picked off the conveyor belt by hand. Always remember to squash the carton before putting it in the bin so it takes up less space, even if you live somewhere where recycling is co-mingled. This is different from plastic bottles (page 105), as squashed cartons will always be sorted as cardboard. You can leave lids on cartons and if it is a small, single drinks carton, push the straw inside.

The aluminium and plastic layers are designed to prevent liquid damaging the cardboard carton outer. This means they take longer to be pulped than standard cardboard.

Drinking glasses and strong glass

Whether it's used in packaging or glassware, all glass seems the same at first glance: you can see through it and there is no obvious difference, other than the colour. However, there are a few different types of glass, some of which are definitely not recyclable.

Drinking glasses and strong glass, which is glass used for cooking, like measuring jugs or plates used in a microwave, fall into the non-recyclable category.

Not that clear

Glass jars and bottles have been designed to be recycled through a melting process and are relatively easy and safe to recycle (page 85). In fact, using recycled glass lowers the melting point of virgin glass, improving the energy efficiency of glass production.

Glass used in kitchenware is around four to six times stronger than glass packaging. This is essential to ensure it is resistant to the sudden changes in temperature that may occur with cooking, reducing the ability for the glass to expand. This heat resistance means it has a higher melting point than glass packaging such as bottles. The different melting points therefore render the two types of glass incompatible, as the difference would create contamination in the recycled glass.

Unfortunately, if a drinking glass or measuring jug breaks, the best thing to do is to put the pieces in the normal waste bin. And remember, it is still made of sharp glass, so wrap it up in heavy paper, like newspaper, first.

Electricals

Obviously there are a lot of products in the category 'electricals', but they broadly fit into fourteen categories (page 73) and are defined as anything with a plug or battery, including components used in electrical products, like ink cartridges.

Electricals can be recycled in two ways: at your local recycling centre, where there is usually a specific area for them, or retailers legally have to take them back when you buy a similar product. So, if you buy a new toaster, the retailer has to take the old one, which you can bring back to the store within twenty-eight days of your new purchase.

Watt a joke!

Electricals contain many valuable metals, which makes them an ideal item for recycling as the work carried out to recycle them is paid for by the value of the materials.

An electrical item can be recycled if it has either a plug, a battery, or a picture of a crossed-out wheelie bin on it (page 129). Remember to take the battery out before – you can drop the battery in a collection point in any store that sells batteries (page 61).

The world of gadgets moves very fast, with consumers upgrading regularly, and second-hand items still hold their value. Reuse is always better than recycling, so if the electrical item still works it is better to find an opportunity for reuse, like donating or selling it, rather than recycling it.

Envelopes and post

Envelopes and post are definitely recyclable and are very likely to be collected from the kerbside.

Confusion usually exists because of the plastic window on some envelopes the different coloured envelopes and the gloss finish of most junk mail.

First-class waste

Brochures, junk mail, flyers and envelopes can all be recycled in the paper or card bin, depending on the colour. They are very suitable for recycling into new paper. The only envelopes that are not recyclable are those padded with bubble wrap which sometimes come from online delivery companies.

Brown envelopes should go in the cardboard bin (as the brown will create flecks on recycled paper) and white envelopes go in the paper bin.

There can be a plastic wrap around junk mail and magazines in newspapers. This film is not readily recycled and should be taken to dedicated collection points. If your local supermarket has a carrier bag recycling bin, you can put this film in there as it is usually made of LDPE, the same material as a carrier bag.

Envelopes with plastic windows are also fine to recycle. Paper mills have ways of dealing with the very small amount of plastic and will remove it before recycling the paper.

Film

It is not just carrier bags (page 142) that are made of flexible plastic film. Film comes in many forms: on the top of many plastic food trays, in bags containing fruit and vegetables, or even in ring pulls (page 180).

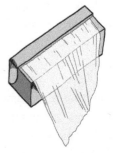

Unfortunately, like carrier bags, due to its lightweight nature and relatively low value, film is not currently widely recycled and must be taken to special collection points.

That's a wrap!

Film is usually made from LDPE, which can be identified from its plastic number, 4 (page 122). At the moment, this is too expensive to collect from the home as it does not weigh very much, making transportation of film from kerbside uneconomical.

Supermarkets have developed carrier bag collection in their stores, and anything made of LDPE can go in those bins. In the future, collecting film from kerbside is likely to be mandated by the government (page 216). This would significantly increase the collection rate of all plastic film.

Examples of things you can put in the carrier bag bin include empty bread bags, the bag in a cereal box and bubble wrap. One of the main products that needs to be checked first is cling film: a lot of manufacturers still use PVC to make this, which cannot be recycled. Some retailers have moved away from PVC in their cling film (usually marked as 'non-PVC'), which can be recycled, if clean.

Increasingly supermarkets are taking a wider range of film, so it is definitely worth checking the bins to see what they are collecting.

Foil

Foil is made of aluminium and, like all clean aluminium packaging, it is 100 per cent recyclable, and its quality will not diminish during the recycling process.

Foil mainly includes aluminium foil that is bought in a roll, instant barbecue trays, metal takeaway containers and trays which are bought to cook food in.

Foiled again

In order to ensure foil is recycled, it is important to make sure it is clean; this means washing off any food residue left on the foil after use. This is really easy to do: just place the foil in the washing-up water when you do the dishes or run it under the tap.

The best way to make sure the foil makes it to a recycling facility is to scrunch it into a ball, ideally larger than a tennis ball. The bigger the ball is, the easier it will be to recycle. If you only have a small bit of aluminium and are worried it will not be recycled, you could push it carefully into an aluminium can. Once the foil arrives at the recycling facility it will get recycled like all aluminium: melted at a high heat and turned into slabs or sheets.

Remember to watch out for metallised plastic film, like crisp packets or chocolate wrappers. This can often look like foil, but cannot be recycled easily and would fail the scrunch test (page 101).

Food waste

Sixteen per cent of food bought is wasted by households each year in the UK.[10] Food waste is a big issue, as wasted food could end up in landfill and release methane, a greenhouse gas. Combine this with the resources required to get food to the consumer,  and wasting any food has a significant environmental impact.

Rotten rubbish

Lots of local authorities now collect food waste, and from 2023 legislation will require separate food waste collections, which is a positive step in ensuring the proper management of wasted food. Food waste is collected in a caddy and it is best to use a compostable bag with the seedling logo (page 131) to line it.

Any leftovers can go into the food waste bin; this includes eggshells, tea bags and both raw and cooked meat or fish. The only things to avoid are packaging, non-food products like nappies (page 170), or liquids which can just go down the sink.

There are two ways for food waste to be managed: composting or anaerobic digestion (page 109). Composting mixes food with garden waste for it to become a fertiliser. Anaerobic digestion uses very small bugs and bacteria to break down the food. The gases released are used to generate electricity, heat or as a fuel.

It is important to avoid food going to landfill by only buying what you need and ensuring any leftovers are disposed of correctly.

Jars

Jars are usually made of the same strength glass as bottles and will therefore melt at the same temperature. They can contain products such as jams, sauces and baby food.

Jars are great at keeping food fresh and are 100 per cent recyclable, even the metal lid. When recycling, it is best to leave the lid on the jar; as with bottles, lids can get lost on their own and putting it back on the jar will reduce contamination.

Sticky is tricky

Glass is a great material for recycling as it can be recycled over and over without losing quality. Unfortunately, glass jars are normally full of sticky things like jams and condiments. It is important to wash these out before recycling as best you can. This is for two reasons: to avoid contamination with other products they may get mixed with, such as paper and card, and to prevent wildlife becoming interested in eating what has been left in there when it is outside in the bin.

Glass packaging can be made with up to 70 per cent recycled glass and adding 'glass cullet' to the manufacturing process reduces the temperature required to melt the glass, saving significant amounts of energy and reducing carbon emissions.

The golden rule – reduce, reuse, recycle (page 98) – really applies to glass jars. It is very common to reuse jars to store things like craft items and homemade preserves. Reusing jars is better than using up energy to turn them back into new jars – just remember to wash them out first.

Kitchenware

Kitchenware covers anything used for cooking, including saucepans, frying pans, mixing bowls, utensils and baking trays.

Like glassware (page 156), in the first instance, kitchenware should not be recycled but instead should be donated to charity or sold. If items are broken and cannot be reused, they should be taken to a civic amenity site (if your local authority accepts them).

Kitchenware and tear

Kitchen items form a broad category, including ceramics and cutlery and drinking glasses. Items will normally be made of metal (pans), ceramics and glass (bowls) or multiple materials (gadgets).

Metal pans can be made from aluminium, steel, iron and copper. In their natural form these metals could usually be melted down and recycled; however, a lot of modern pans are covered with chemicals that stop food sticking to them. In order to be able to recycle these pans, the chemical would have to be removed first, which most recyclers are not able to do.

With kitchen gadgets and utensils, the best thing to do is first check if they have a plug or use batteries. If that is the case, these are electricals (page 157), and can be dismantled and recycled. Otherwise, items like whisks, tin openers and wooden spoons can be taken to a civic amenity site (always check locally first), or thrown in the normal bin if your council doesn't accept them.

Lightbulbs

Lightbulbs have changed a lot over the years, from older style, inefficient incandescent bulbs, which have been phased out, to newer style 'energy efficient' LED bulbs, which are better for the environment, as they use around 75 per cent less energy and last a lot longer. Newer LED bulbs can be recycled, whereas older filament bulbs cannot.

Lighten up

Older style bulbs were banned from sale in 2012, so any bulb bought before then needs checking as it may not be recyclable. Incandescent bulbs look like the one in the picture opposite; they are glass bulbs with metal inside them. These are not recyclable and should be put in the normal rubbish bin. Do not put them in your glass bin: they contain metal, and the glass is not the same as the type that makes up bottles and jars. Since these older bulbs are banned, when lights are changed in your house, they will be replaced by newer recyclable bulbs.

Newer style, energy efficient LED bulbs are more likely to last longer and use less electricity. If the bulb does not contain visible metal in the glass, then it can be recycled.

In order to recycle them, they can either be taken to the local recycling centre, or there are special collection points in large supermarkets. Next time you are in a large supermarket look for a lightbulb collection point, then you will know where to take your bulbs for recycling.

Magazines and newspapers

Magazines and newspapers can be recycled (but be sure to watch out for any plastic wrap around them or free gifts that may come attached). Paper is a valuable material and recycles well, but the quality of recycled paper is affected  by contamination with food waste and liquids.

Start shredding the news!

Magazines and newspapers, like most paper, are recycled through a process called 'pulping'. They are mixed with water and chemicals and then rolled out into pulp and dried to remove the water (page 84). This process creates recycled paper, like the paper used to make this book.

Each time paper is recycled, the fibres which make it up will get shorter, preventing them from sticking together and being useful. To combat this, virgin paper is added to the mix to introduce new and longer fibres. Magazines and newspapers tend to use lower quality paper, which can have a higher recycled content.

Magazines and newspapers can be wrapped in a plastic film, so, as with junk mail, this needs removing and either putting in the normal bin or taking to carrier bag collection points in supermarkets, which will recycle this kind of film (page 59). The free gifts and toys that come with some magazines should be removed before recycling as these are unlikely to be recyclable (page 185).

Mattresses

We've all seen the adverts for new mattresses sold online: if you don't like it, you can return it within a hundred days. This approach to sales has created a recycling challenge, with more consumers changing their mattresses more frequently and only around 20 per cent of mattresses being recycled,[11] and materials therefore being lost to landfill or incineration.

How do we sleep at night?

Over 7 million mattresses are thrown away each year. Unfortunately, the make-up of materials in a mattress makes them a real challenge to recycle.

The combination of springs, foam and fabric would damage machinery in a traditional recycling plant. This means mattresses need to go to specialist recycling centres, of which there are not many, leading to low recycling rates.

Mattresses are shredded in powerful machines, with blades that need changing regularly. The materials are separated with magnets and air. The foam can be used to make up carpet padding, the fabrics can be reused and the steel springs can be melted down and used in multiple applications.

If your mattress is in good condition you should sell or donate it, so it is not wasted.

Mattresses at the end of their life may be taken at a household recycling centre, but be sure to check your local one accepts them. Alternatively, you can pay a small fee to have a company, or sometimes the council, collect it, for it to be taken away for recycling.

Milk bottles

All milk bottles are recyclable and are usually made from HDPE with a plastic lid, or reusable glass with a foil lid.

Got milk... bottles?

Plastic milk bottles are bought from the retailer, whereas glass milk bottles are normally delivered directly to households.

The milkman was a regular sight in the UK in the twentieth century, but was overtaken by a rise in supermarket shopping. However, due to a conscious effort to avoid plastic there has been a significant resurgence in glass milk bottles. There is some debate over which is actually better: heavy and reusable glass, or light and recyclable plastic (page 104).

Plastic milk bottles are made of HDPE, identified by their plastic number 2 (page 122). They are widely recycled and should be placed in the plastic recycling bin when finished; make sure there is no liquid in the bottle, and as always put the lid back on. Some supermarkets now use plastic milk bottle lids that are a lighter colour so they can be sorted easily.

Glass bottles should be returned to the milk company that delivered them. This is really easy: just put them on your doorstep when they come to deliver new milk and they will be taken away. In theory, a glass milk bottle could be reused fifty times before reaching the end of its life, significantly reducing the environmental impact of creating the bottle in the first place. However, in practice bottles are generally not reused as often as this before being recycled.

Mirrors

Mirrors are primarily made of glass and so you would be forgiven for thinking they can be recycled if broken.

Unfortunately, this is not the case and, as well as bringing a bit of bad luck, a broken mirror will have to go safely into the normal waste bin.

Mirror, mirror had a fall

Mirrors are essentially a giant piece of glass which has been sprayed with a thin layer of metal. It is this sheet of metal that prevents mirrors from being recycled as it is not easily removable.

Like drinking glasses, mirrors are made from a glass which has a different melting point to the glass in bottles or jars. This means that they will not recycle at the same temperature and could potentially contaminate the glass cullet that recyclers create.

If your mirror has broken, remember it is still made of glass and could be quite dangerous to rubbish collectors as its sharp edges will poke through bin liners. Make sure you wrap the shards in heavy paper, like newspaper, before putting them in the bin. Do not worry if this means the paper is not being recycled, as it is definitely better to be safe than green on this occasion.

As with all things, reuse is better than recycling. Mirror frames can make great photo frames, and the internet is full of ideas and craft projects to upcycle a broken mirror.

Mobile phones

Mobile phones technically fit into the 'electricals' section and are classed in the IT and telecommunications equipment category (page 73). This means they can be recycled; however, unlike larger bulky electricals,  they are small enough to be recycled by sending them in the post.

iRecycle

As new mobile phones come out each year, many end up reaching the end of their life very quickly, with most consumers upgrading their phone every two years. This makes recycling them important as the chemicals and plastics can damage the environment if sent to landfill and the valuable metals they contain could be lost.

Luckily, mobile phones are worth money as they contain precious metals like gold and silver. Typically most people get rid of phones by selling them on or donating them to charity. It is the components or reusability of the phone that gives it this value, something we don't always appreciate. In order to sell your phone safely it is important to remove personal data first, which can be done in the phone's settings.

Before a phone is recycled it will be assessed by the company that has bought it for reselling, either by fixing the phone or by cleaning it up. If a phone cannot be reused it will be recycled. Recycling the phone involves assessing how the phone was put together and working backwards, removing parts piece by piece, a process mobile phone companies should make easier.

Nappies

Nappies always find themselves on lists of things that people recycle when they should not (referred to as 'wishcycling', page 196).

They make up a large proportion of waste from families with babies

and news coverage about them being recycled leads to confusion. At the moment, they will usually belong in the waste bin.

Very absorbing

Disposable nappies are a huge problem, making up around 3 per cent of all household waste, with each baby potentially using up to 6,000 disposable nappies before they are potty trained![12]

It is possible to recycle disposable nappies, and they contain valuable plastics. To start with, the nappies are opened up and sterilised using steam. Once the nappies have been sterilised, they are shredded and sorted into their component parts. This process is entirely automated. Unfortunately, as with many products, the issue is collection: very few local councils currently collect nappies for recycling, contamination being a major concern, and so they should be placed in the normal waste bin.

Instead of using a disposable nappy you could buy reusable and washable cloth alternatives. Of course the life of a reusable nappy is not carbon free – it will generate carbon emissions from the intensity of the washing and drying process. This can be reduced by including the reusable nappy in a full load of washing, on an efficient setting, and letting them dry on a line, rather than in a dryer.

Paper

Paper comes in many forms, including white office paper, recycled newspaper, shredded documents, brown envelopes, labels and wallpaper.

Most paper can be recycled and will be turned into pulp. However, there are specific types that are too difficult to recycle easily and should go in the normal bin.

Tear-able waste

This book has already covered the recyclability of paper (page 84). However, contamination and a potential reduction in quality each time it is recycled means that in fact paper isn't always recyclable.

If paper has been stained with paint, dyes or glue it is not easily recycled, as the recycler will not be able to remove these contaminants from the paper. For this reason, things like wallpaper cannot be easily recycled as it usually contains dyes that colour the paper and glue that sticks it to the wall, so it should just be thrown in the normal bin. This also applies to any paper with paint, dirt or grease on it.

Brown paper should be put in with the cardboard, because the brown colouring would cause brown specks to appear on white paper if they were recycled together.

White paper and shredded paper can be recycled and should be placed in the paper bin. Labels stuck to packaging, like cardboard boxes, should just be left on the packaging.

Pill packets

Blister packs, like those containing pills, gum or contact lenses, are usually made of plastic, normally PVC, with a foil layer stuck to the top to allow the product to be removed

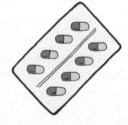

easily. This combination of materials makes them very difficult to recycle; PVC is not widely recycled, and the composite nature of the aluminium and plastic means they belong in your normal rubbish bin.

By gum!

Blister packs are designed to keep small items protected from external elements and therefore safe from contamination. This is of particular importance with healthcare and food items, like gum. They are also helpful for ensuring one dose can be taken at a time, rather than exposing additional doses of medication to the elements every time a tub is opened.

To create a blister pack, the plastic is heated until it becomes pliable, and a mould forms the cavity before the product and film are added.

They also ensure products cannot be tampered with; it is obvious if someone has removed a pill or attempted to switch it for something unexpected, unlike with equivalent packaging, such as a rigid plastic bottle.

Anything else that comes with the packaging, for example the paper instructions or cardboard box, can be recycled and should be placed in the correct bin.

Pizza boxes

Made of cardboard, and delivered to your home with hot food inside, takeaway pizza boxes are recyclable... most of the time.

There have been many news stories detailing how pizza boxes are destined for the normal waste bin. However, like most things in this book they do not need to be, with a little bit of work.

Greased frightening!

The problem with pizza boxes is the pizza inside. It is made of cheese and dough, both of which can produce a lot of grease. When the cardboard gets to the recycler it is shredded and mixed with water and chemicals in order to produce pulp. Some contaminants will be removed, but grease will not. This means that a thin layer of grease on the box could contaminate everything else.

In order to recycle a pizza box, the first thing to do is empty the box – eat up! The box should be completely empty – do not leave the crusts or empty dip containers in there, and remove any food residue.

The next step is to look at how much grease there is. If the whole bottom of the box is covered, it is not recyclable. If there are just splashes of grease, put it in the recycling bin. If in doubt, cut around the greasy parts and throw those in the normal bin. Most pizza boxes have a perforation between the lid and the base, and the lid will rarely have grease on it, so an alternative solution is to separate the two parts and only recycle the lid.

Polystyrene

Polystyrene comes in many forms and is difficult to recycle. Expanded polystyrene is rarely collected from the household.

Rigid polystyrene, like yoghurt pots, may be collected, but unfortunately is unlikely to get recycled.

An expanded problem

Polystyrene comes in forms as diverse as DVD cases (remember those?), yoghurt pots (although these are less likely to be polystyrene as covered next), plastic cutlery and packing materials. You can identify it from its plastic number 6 (page 122).

Polystyrene is most common in a form known as 'expanded polystyrene'. This is the type that makes up takeaway containers, like the one pictured opposite, or the pieces of white 'peanuts' that are used in the packaging industry. It tends to get everywhere when you unpack a new TV. It is so widely used because it is cheap to create and does not weigh very much (in fact it can be up to 95 per cent air). It is also these properties that make it very difficult to recycle economically as it is hard to collect, and it is cheaper to create it from scratch.

Expanded polystyrene is so light it floats on water and can look like food to wildlife, which can not only be really harmful to their health but also contributes to plastic entering the food chain. The only safe way of getting rid of this type of polystyrene is for it to be taken to special incinerators where it is burned, and this may happen if you put it in your normal waste bin. Some councils are looking to introduce polystyrene collections at civic amenity sites, though, so it is definitely worth checking local websites.

Pots, tubs and trays

Pots, tubs and trays encompass a huge proportion of plastic packaging, including yoghurt pots, ready-meal trays, fruit punnets and margarine tubs.

In 2017, the category was deemed to have improved its collection infrastructure enough to be awarded the 'recycle' label from OPRL (page 133), which means over 75 per cent of local authorities collect them for recycling.

Strawberry pun-net

Pots, tubs and trays are usually made out of PET and PP, which are relatively easy to recycle. Yoghurt pots are sometimes made with rigid polystyrene, which may be collected but is unlikely to be recycled. Most yoghurt manufacturers are now moving away from polystyrene, so it is worth checking by looking for the plastic number (page 122) and the OPRL 'recycle' label.

Unfortunately, black pots, tubs and trays cannot be recycled as the sorting machines cannot identify them using near infrared (page 108). However, a lot of companies are now moving away from carbon black or may begin adding a tag for the sorting machines to read (page 235).

Since over 75 per cent of local authorities now collect pots, tubs and trays, it is worth checking to see if they can be recycled in your area, as chances are they can be.

Remember to remove flexible plastic film lids from trays and pots and ensure the packaging has been washed out before recycling, to avoid contamination.

Pouches

Pouches are usually made of metallised plastic film, which is used to protect food and drink. They can contain various items, including pet food, coffee, baby food and even drinks (normally with a straw to pierce the container).

Cat-astrophe

Pouches have become a popular container for food items and help to increase shelf life in the store. They are extremely lightweight, compared to a can for example, so are easy to transport. They have become so popular that over 10 billion are sold each year.[13]

Unfortunately, these pouches are not easy to recycle. The thin layer of aluminium that's stuck to the layers of plastic prevents it from being recyclable, but also makes it much lighter than a tin can, which would be the alternative packaging.

Some companies have proved that pouches can be recycled through a pyrolysis process (page 90). Unfortunately, at the moment this process is expensive and difficult to scale. As metallised plastic film is difficult to recycle, pouches are not regularly collected from the home, which means they need to go in the normal waste bin.

To combat this lack of recycling, manufacturers have worked

to create pouches which do not have the layer of aluminium and are made of only plastic (page 227), which could be easier to recycle and would fall into the plastic film category, meaning kerbside collection is likely in the future.

Sachets

Sachets can be made of plastic, paper, foil or metallised plastic film. Plastic and foil sachets are often used to hold things like dishwasher tablets. Paper is used to hold sugar and metallised plastic film sachets are used to hold condiments, like tomato ketchup, mayonnaise or mustard for single servings.

Sachet away!

The most common use for sachets is in the catering industry or in hotels to serve condiments or toiletries. Normally this type of sachet is made from metallised plastic film, which is too small to capture for recycling and heavily contaminated with food after use. This makes them almost impossible to collect, transport and recycle at scale and so sachets will just go in the bin, destined for incineration or landfill.

This has led to many environmental groups calling for sachets to be banned; after all, it is easy to provide a bottle which contains more servings, dramatically reducing the environmental impact of the equivalent product. As with all issues related to recycling, it is a nuanced problem, as in reality the weight of a sachet is lighter than a bottle on a per serving basis, similar to a pouch versus a can. However, its size and lack of recyclability nonetheless makes it one to avoid.

It is a surprise that sachets were not included in the list of single-use plastics that have been banned, as they are single use by design. This seems to be an omission that may be rectified in future legislation.

Sandwich boxes

In the UK we get through over 11 billion pre-packaged sandwiches per year, or 200 per adult![14] Sandwich boxes are deemed to be recyclable by OPRL (page 133) and display the 'recycle' label, which indicates that over  75 per cent of councils collect them. However, they are made of cardboard with a plastic liner, and many say they cannot be recycled, including chef Hugh Fearnley-Whittingstall in his 2020 BBC programme, 'War on Waste', so what is the truth?

BLT (Box Lined Terribly)

Sandwich boxes consist of a card outer casing, usually with a plastic window and a plastic liner to keep the food fresh. The key to whether or not a sandwich box can be recycled is the amount of plastic included. Paper recycling mills can tolerate a certain amount of plastic in the process. This means if the sandwich box has a 'recycle' label on it, you should put it in the cardboard bin. However, if you can remove the plastic lining first that really helps.

The maximum percentage of plastic that sandwich boxes may contain while still qualifying as recyclable is 15 per cent by weight, which will be reduced to 10 per cent by 2023. Any sandwich box that contains a proportion of plastic greater than these percentages will be classed as 'not recycled' by OPRL.

Some sandwich boxes are now manufactured with a tab on the inside which, when pulled, may remove the plastic lining that is attached using peelable glue. Having tried this on multiple brands' boxes, this is far easier said than done and it tends to break up the boxes and leave a mess. Once this solution has been refined, this could be a good way to remove the plastic.

Six-pack rings

Used to hold multipacks of cans, plastic rings are an environmental problem, consistently shown as a hazard to wildlife that get stuck in them. However, they are also the best example of how innovation can be used to reduce packaging.

Glue can(s) work

Before plastic versions came along, cans were held together using paper and metal. Since the 1960s, plastic variants have become commonplace, with convenience taking priority over environmental considerations.

They have developed a reputation alongside products including straws as a scourge of the ocean. Compared to other rubbish such as fishing equipment, six-pack rings constitute a relatively small fraction of ocean waste. However, images of wildlife trapped in them have highlighted this unnecessary problem, considering cans could simply be bought loose.

Rings are normally made of LDPE and therefore could be recycled in the same way as film, but this is unlikely to happen given their size and the need to take the rings to a supermarket to recycle them.

For this reason, companies have developed innovative alternatives, from edible versions made of products such as barley, to holders made of cardboard. However, considering the fact that reducing is better than recycling, arguably the most successful replacement is the dab of glue some manufacturers are using to stick the cans together. This residue will be burned off in the metal recycling process.

Soap, shampoo and shower gel bottles

Bottles found in the bathroom containing soap, shampoo and shower gel are often thrown away and not recycled as they tend to be placed in bins in the bathroom, where it is easy to forget to separate the recyclables.

Like a lot of rigid plastics, this material is valuable and should be sorted into the plastic recycling bin when finished with.

Soaper-duper!

All the bottles found in a bathroom just need to be emptied, rinsed out and then recycled in the plastic bin. Bottles in a bathroom are usually made of PET, HDPE or PP (page 122). These rigid plastics are likely to be collected by your local authority and are highly recyclable.

The only problem with soap bottles is the mechanism you push down to get the soap. These dispensers have a metal spring in them to ensure the lid bounces back up. This spring means these lids cannot be recycled. If you have to push the lid down, just throw it in the normal bin before recycling the bottle. These lids are different from triggers which you squeeze, as these are less likely to contain metal and can usually be recycled.

Most liquid soap companies will produce a refill bottle without the pump so that you can reuse the old pump. It is definitely worth reusing the pump where possible due to its lack of recyclability.

Sticky tape

Every time a cardboard box is delivered to your house it is likely to have tape attached to seal it. Although a key component in packaging, it will regularly end up in your bin as, unfortunately, sticky tape cannot be recycled. It is best to remove it before recycling things like boxes or wrapping paper, although it will not necessarily affect its recyclability, due to its relatively low weight.

Time to tape responsibility

Sticky tape is made from plastic film (normally PP) and it is therefore difficult to recycle. Some tape being left behind is inevitable, but it is helpful to remove as much as you can before recycling to ensure the quality of the recycled paper is as good as possible.

When paper-based packaging (such as a cardboard box) enters the recycling process, it is mixed with water and broken down to pulp (page 84). When this happens, contaminants like plastic tape will separate from the fibres and will be removed from the mix. Unfortunately, too much tape could cause problems to the sorting machinery, which is why it is best to remove it first.

Even other types of tape that are not plastic, like masking tape, cannot be recycled due to the adhesive layer.

The core of adhesive tape, whether cardboard or plastic, is recyclable and should be placed in the correct bin.

Straws

Disposable straws, whether paper or plastic, are the definition of single use. A staggering 8.5 billion straws are thrown away each year, although this number has significantly reduced following a ban on plastic straws (page 207).[15]

Unfortunately, both paper and plastic straws are too small to be successfully recycled.

Grasping at straws

In the UK, plastic straws have been banned from sale since October 2020, although there are limited exceptions for people who need them, for example for medical reasons. Alternative materials that are still allowed include paper and reusable straws.

While this is a positive move in the effort to reduce single-use plastic, the inevitable replacement with paper straws creates other environmental issues. The creation of paper has a higher environmental impact than plastic and these straws tend to break up when used and will be too contaminated to ever be recycled, even if collected.

As with most single-use items that are only used for a short amount of time, the best thing to do is avoid them. Luckily, in most cases straws are not needed, so just asking not to have one is the simplest way to do your bit to help.

Reusable straws are becoming more common; these can be made of metal, glass or hard plastic and can be washed at home and reused.

Tissues, toilet roll and kitchen roll

Paper tissues, toilet roll and kitchen roll are not recyclable.

Unfortunately, the paper is not high quality enough to go through the paper recycling process, due to the shortening of the fibres that make up paper each time it is recycled.

Not to be sniffed at!

It takes around 384 trees to make the toilet paper one person uses in their lifetime![16] None of this enters the recycling system and will usually be flushed down the toilet instead (note that facial tissues and kitchen towels should be thrown away in the bin). When buying toilet or kitchen rolls, look for ones made from recycled paper; since they get thrown away anyway, you can help by buying rolls made with recycled material.

Tissues, toilet roll and kitchen roll will always be used in a manner that contaminates them (this is as polite as it gets!), which makes them unsuitable for the recycling process.

Fortunately, kitchen and toilet roll are wrapped on a cardboard core that can be recycled in the cardboard bin. Any boxes tissues come in can also be recycled in the same way.

Some local authorities that collect food waste will let you put small amounts of used kitchen paper in the food waste bin. This will be mentioned on your council website, so have a look and see if you are able to.

Toys and games

'Toys and games' is a category that covers a huge number of items, which cannot all be assessed here. However, a common feature is that most toys will be made of rigid plastic.

Toys and games will not usually be collected at the kerbside, but they can be donated or may be able to go to your local recycling centre, although you will need to check.

Playtime is over

Toys and games can be made of many materials, such as cardboard jigsaws, plastic dolls and wooden blocks. The variety of materials they are usually formed with and their bright colours make it unlikely that they'll be collected from the household.

There are separate rules for electric toys, which have their own category under waste electrical legislation (page 73) and so can be taken back to a store for recycling when a new and comparable toy is being purchased (page 60). These electrical toys are likely to contain a battery which should be removed and recycled through a battery collection point (page 61).

Some local recycling centres will accept toys and games, so it is definitely worth checking local websites to see if yours will.

Alternatively, if the toys are in working order, they can be donated to a charity shop or sold at a car boot sale or on second-hand sale websites.

Tubes

Squeezable tubes contain a variety of materials and can be found in the bathroom as a tube of toothpaste, or in the kitchen with products like tomato purée.

Just as the material they are made from varies, so too does their ability to be recycled.

A bit of a squeeze

Most tubes you use can be divided into two categories: plastic, like the tubes that contain toothpaste, or metal, containing food like tomato purée.

Plastic tubes that you squeeze cannot be recycled; the residue left inside them, which is impossible to clear, would contaminate other plastic recycling. These tubes tend to have a metal layer inside to keep things like toothpaste fresh, which also affects their recyclability. This does not just apply to toothpaste tubes but also to other potentially contaminated products like mastic tubes used in DIY. Some tubes with a pump action (for example, toothpaste in hard plastic tubes) can be recycled as they are made of harder plastic and are therefore treated like a pot, tub or tray (page 176), so if your local authority collects these, they can probably be recycled.

Tubes made of metal, like the ones containing tomato purée, can be recycled and belong in the aluminium bin. Give them a rinse first to reduce possible contamination. The plastic lid is likely to get removed in the recycling process, but, as with many other products, it is fine to leave this on the tube when recycling.

Tyres

Tyres are recyclable and are usually returned when you purchase new ones. In fact, around 95 per cent of tyres in Europe are recycled.[17] Despite this high recycling rate, tyres are also one of the largest sources of microplastics found in the ocean.[18]

Wheely bad

Tyres are usually disposed of for a small charge when they are changed. Tyres will not degrade in a landfill and if burned they produce toxic fumes and thick black smoke, so environmentally it is better if they are reused in applications where rubber is required.

Like lots of plastic, a tyre will usually be shredded at the end of its life. The shreds are most likely to be used as a source of fuel. They can be used as an energy source for kilns, as they burn hot for a long time. They are also used in the building of pavements, running tracks and roads. Old tyres may also be repurposed into things like swings, playground equipment and plant pots.

Pyrolysis can be used to turn the tyres into a clean fuel source by burning them in an oxygen-free environment and removing vaporised gases that can then be used as a fuel source.

Through normal use and contact with the road, the average tyre will release 4 kilograms of microplastics (page 205) in its lifetime;[19] these then get blown around, eventually settling in the oceans. Tyres are a major source of microplastics, and this is a problem that could get worse as consumers move towards typically heavier electric cars.

Wet wipes

Wet wipes, including cosmetic wipes, baby wipes and household cleaning wipes, cannot be recycled and should not normally be flushed down the toilet either. They were developed in the 1970s and are made of a  combination of materials squished together into sheets.

Unflushable

A lot of wet wipes are marketed as 'flushable', but with a few exceptions, this is not true. Wet wipes, usually made of a mix of cotton and plastic, will not shrink in water in the way that toilet paper does.

Wet wipes can cause a lot of environmental damage as they can block drains and pipes, causing other items, like food and oil, to stack up and stick to them, clogging the pipes further.

The complex mix of fibres and plastics also means wet wipes cannot be recycled, as it is not possible for recyclers to separate these materials. This means they should just be put in the normal waste bin, where they are likely to be incinerated or end up in landfill. Unfortunately, even in landfill they will not break down. It is probably best to stick to paper-based products, like tissues, toilet or kitchen roll (page 184), or use reusable alternatives to wet wipes.

Due to the confusion around whether or not wet wipes are flushable, a UK standard was developed by Water UK in 2019. This means that some wipes are able to be flushed; these are marked with a 'Fine to Flush' logo. These wipes meet rigorous testing standards and will not contribute to clogged drains.

Wrapping paper

Wrapping paper definitely falls into the 'it depends' category. A lot of it can be recycled, but some of it cannot, and unfortunately it definitely depends on where you live, and the type of paper mill your local authority sends its waste to.

Some councils will collect it from your house, some will ask you to take it to the local recycling centre and some will not collect it at all. This is mainly because of the different technologies at paper mills.

All wrapped up

Wrapping paper is designed to look attractive: it is often colourful and may be decorated with non-paper items, like glitter, which are microplastics (page 205). These cannot be recycled easily.

Luckily, there is a really easy way to tell whether wrapping paper is recyclable: the scrunch test (page 101). Just scrunch the wrapping paper into a ball: if it holds its shape as a ball, it can be recycled; if it opens back up again it cannot be recycled. This is due to the proportion of plastic used in the wrapping paper. The more plastic there is, the more likely the wrapping paper will open back up and the less likely it can be recycled.

Consider if the paper can first be reused, perhaps to wrap something smaller, or in arts and crafts projects. If it cannot, always remember to remove sticky tape, bows and decorations before you recycle wrapping paper. As the name suggests, it will go in the paper bin.

There you have it, fifty of the most common items you will find around your home have now been assessed. It might be worth keeping a copy of this book next to your bin should you need a reminder on recycling day! Remember though, councils collect and recycle differently, so the best practice documented in this book needs checking against council websites.

9

The Rubbish Problems

There is rarely a day when the media is not writing something about recycling, invariably focused on the problems and challenges, with an obsessive focus on plastic. Let's look at some of those problems in more detail and see if things really are as bad as they might feel.

Is packaging actually bad?

By this point we have reviewed the recyclability of a lot of packaging and considered the creation and disposal of several material types. You might be asking the question, 'Is packaging good or bad?' Should we simply be avoiding plastic or cutting packaging out altogether?

Pain in the glass

Remember, there is no perfect packaging. The more often a product is collected and recycled, the more difficult the raw material or packaging will be to obtain in the first place (page 11). As an example, cans and glass bottles have high recycling rates as there is a higher energy cost associated with their creation and disposal than that of a plastic bottle.

You may have heard of people looking to lead a 'plastic-free' life. While the motive for doing this is understandable, adopting this approach will not necessarily lead to a lifestyle with a lower carbon footprint. This is because alternatives to plastic often require a far greater amount of energy to manufacture, and can release gases that contribute to global warming when they break down.

As a lightweight, cheap and durable material, plastic certainly has a place. However, the proliferation of it as a material has led to the development of pointless packaging. Sachets instead of bottles, straws for non-medical use and magazines wrapped in film should all be eliminated wherever possible. Then there are the plastics that are not economical to recycle, such as polystyrene and PVC, which should be replaced with recyclable alternatives, if possible.

It is worth remembering that pointless packaging can be made from all kinds of materials, not just plastic. Have you ever thought about why toothpaste tubes are sold in cardboard boxes? These boxes are completely unnecessary beyond improving presentation on a shelf, and are thrown away immediately once the product has been opened. We should be challenging all forms of packaging that do not serve a purpose, regardless of the material they are made from.

Where things get trickier is in relation to food packaging. Consumers often ask why plastic film is needed around fruit and vegetables, forgetting that long supply chains require food to be fresh for a longer period of time than nature would allow without packaging. How long does bread usually stay fresh without a bread bag? Hours, at best. And we could not have safe milk without packaging.

When we take into account the energy expended throughout the entire supply chain, as well as the gas emissions from waste food breaking down, the environmental implications of food waste are more significant than the implications of packaging waste. Without packaging, food waste in supermarkets and in households would rise significantly. Amazingly, on average packaging only makes up around 5 per cent of the environmental impact of a food product.[1]

To reduce your packaging or waste footprint, one of the things you must do is ensure you review the whole lifecycle of a product. Consumers seeking to avoid plastic around products like toilet roll may order an alternative paper-wrapped product online which is manufactured and shipped from China. In reality, the environmental footprint of this, is far worse than a locally sourced recycled toilet roll wrapped in plastic.

Plastic has been demonised, and other packaging materials promoted instead. However, the environmental credentials of these alternatives are rarely interrogated to the same extent as the plastic they have replaced. Ask yourself: would you feel better drinking from a can or a plastic bottle? If it is the can, is this simply because plastic feels bad?

Our packaging use today is the result of a drive for convenience; we expect to be able to buy certain foods every day, regardless of the season, which means they have to come from all over the world, and therefore need to be packaged. If we shop more locally, and change our consumption habits, we would not need packaging in anywhere near the same volumes.

Assuming convenience in society is important, we should eliminate pointless packaging; review the entire lifecycle of packaging that is essential to extending a product's life; avoid making knee-jerk decisions without thinking critically; and consider how we can help ensure packaging is collected and recycled to improve its overall sustainability journey.

What is landfill?

The definition of landfill is the burying of waste material in pits. Most people think of a landfill as a hole in the ground filled with waste; however, an actual landfill is more complicated than that.

Pit of despair

Landfills are the oldest form of waste management; there is evidence of landfills from as far back as 3000 BC.[2] They remain a common method of waste disposal around the world, but their popularity and political support are waning. Waste in a landfill breaks down very slowly and remains a problem across generations, pumping toxins into the environment, and this is why landfills belong on the lowest rung of the waste hierarchy (page 96).

Before waste is added to a landfill, layers of plastic lining, fabric, soil and gravel are put down to seal the base. This prevents any residual liquid from seeping into the soil and contaminating the area. Waste is added in sections at a time and machines roll over the sections to compact the waste, making it as thin as possible. Once a layer of waste has been added, it is covered with soil to reduce odours.

When the waste breaks down, methane gas will be released, a major component of natural gas. These gases are extracted and used to create electricity or other forms of energy.

Once a section of landfill is full, it will be topped with a plastic cover. The land above the waste is then restored, using linings and soil.

Since 1996, a landfill tax has been imposed in the UK, which is generally increased by the government annually. This tax is seen as one of the most important drivers for increasing recycling and reducing landfill. It is paid by businesses sending waste to landfill, essentially increasing the incentive to recycle or incinerate.

What is 'wishcycling'?

Also known as aspirational recycling, 'wishcycling' is when you put something in a recycling bin that cannot be recycled, in the hope that it somehow will be. This can be worse than throwing something away that could be recycled, due to the impact of contamination on other products that can be recycled.

When you wish upon a jar

Over 80 per cent of people admit to having tried to recycle something that cannot be recycled.[3] This is a figure that appears to be increasing, as people respond to the greater focus in the media on the damage that packaging can have on the environment.

In theory, this sounds good: better to be over-recycling than not recycling enough. However, this is not the case.

Major contaminants, such as drinking glasses, kitchenware, clothes and nappies, can affect a whole load of recycling, meaning it could all be sent for incineration or landfill. In this situation it is actually better to put products into the normal waste bin if you are not sure, in order to protect the quality of recycling collected.

Plastic films and flexibles, such as carrier bags or vegetable bags, make up the items most 'wishcycled'. This is perhaps unsurprising due to the amount of flexible plastic we get through, and given that it is a plastic type (LDPE) that can often be returned to most large supermarkets, people may mistakenly believe it will be collected from kerbside.

Other common culprits include batteries, toothpaste tubes, plastic toys and cartons being put in the recycling in areas where the council does not collect them. While these items are a problem, they are not seen as major contaminants, as they are recyclable and unlikely to contaminate or damage other packaging.

What is energy from waste?

Waste can be converted into energy, such as electricity and heat, through burning. It is important to remember this in the context of the waste hierarchy (page 96): energy from waste, referred to as 'recovery', sits below reduce, reuse  and recycle, but above landfill. However, with an increasingly renewable electricity supply, questions arise as to whether energy from waste should now sit below landfill in the waste hierarchy.

Burn wastey burn

Energy from waste is produced by burning residual material, the waste that is left over after sorting which cannot be recycled. This ensures the waste hierarchy is applied, and material that can be recycled is not burned. The introduction of the landfill tax (page 195) led to the development of a new generation of energy from waste plants, focused on low carbon energy.

There are methods of creating energy from waste that are more complex, such as 'gasification' and 'pyrolysis', although these are less common. The most common way of creating energy is to harness the heat created from the burning of waste to boil water, creating steam. The steam is then supplied to a steam turbine to create electricity and heat. The waste gases created are cleaned before emission to ensure they meet strict environmental standards.

As the UK moves to a decarbonised future, with more energy generated from renewable sources, incinerating waste may cease to be seen as a 'low carbon' energy source, as it could be more carbon-intensive than most of the other suppliers to the grid. Also, the composition of our waste is changing: the more carbon-based packaging in the mix, such as plastic, the higher the carbon output. This means in the future it might actually be better for waste to go to landfill than to incineration, an eventuality that the waste hierarchy never considered.

Why do batteries catch fire?

Astonishingly, over a third of all fires at recycling plants are caused by lithium ion batteries, and this statistic is probably not the full picture as the cause of most fires is never reported.[4] Battery fires are a rising problem, as so many of our gadgets nowadays require batteries. So why do they catch fire and how can we help?

Fire power

Lithium ion batteries are generally safe, lightweight and store lots of power, but the millions sold worldwide each year will inevitably lead to some failures, particularly in situations where they could be crushed or punctured, as they can be in a recycling facility. Most combustible items, such as gas tanks, fireworks or even aerosol cans are easy to spot and remove. Conversely, batteries are small and sometimes integrated into products so they can be impossible to pick out in the sorting process.

In a lithium ion battery, the electrodes are separated by a thin piece of plastic. If damaged, the plastic layer can fail, causing the electrodes to touch and ignite. Batteries can trigger a 'thermal runaway event', which means one heat-releasing process can trigger other processes, leading to an uncontrollable increase in temperature and potentially a fire. This fire can then spread to more combustible material, such as paper, cardboard or plastic, and then get carried around a recycling plant on the conveyor belts, spreading the fire and damaging machinery as it goes. This is not just a problem in recycling facilities but also in the trucks transporting the waste to these facilities, as the compacting of waste in trucks can also lead to batteries being damaged and catching fire.

So to avoid fires and to protect valuable resources, do not throw batteries away, but make sure you recycle them responsibly.

What is National Sword?

National Sword was the name given to a policy implemented by the Chinese government to reduce the amount of waste being imported from other countries. Arguably, this policy has created the biggest shift in attitudes towards waste management globally, since its introduction in 2018. It made companies realise their waste may not always have somewhere to go, something that had previously been taken for granted.

Made in China

Historically, waste managers in the UK have relied heavily on China as a destination for waste. Roughly two-thirds of the UK's paper and cardboard and over half of our waste plastic is exported to other countries for recycling.[5]

China's historical desire to import waste stems from its reputation as a manufacturer. As a significant producer of goods that are used worldwide, China had a need to recover material through imports for its future manufacturing. However, as the country has grown and developed better domestic recycling of waste from its own population, China no longer needs to import as many materials. This led to the implementation of National Sword in March 2018, which put strict quality controls on the waste China would accept, effectively halting the export of low-grade plastic, unsorted paper and contaminated material, particularly paper, cardboard and plastic. The acceptable level of contamination dropped from between 5 and 10 per cent, to just 0.5 per cent.

This had a major impact on the global waste industry, as many countries relied on China as a destination for their waste. Exports subsequently increased to other countries like Malaysia, Turkey and Indonesia, which were willing to fill the gap left by the Chinese ban.

China had built a very established waste management industry from years of importing waste. This created fears that the countries that came forward to fill the gap would not have similar controls, which could lead to an increase in uncontrolled waste management, and therefore an increase in issues like ocean plastics (page 208) due to plastic escaping the recycling process.

Indeed, when the policy was introduced, the value of materials dropped significantly as there was a backlog of waste in many countries that just wanted to get rid of it. Many of the countries that stepped forward to import waste are now implementing their own restrictions, having received far too much waste at once, and are sending back highly contaminated waste to the country of origin.

Exporting waste has long been a problem, and a significant proportion of UK cardboard and plastic is sent overseas. This led to the Environment Agency restricting mixed plastic being exported outside of the OECD (Organisation for Economic Co-operation and Development – a group of 38 countries) from 1 January 2021. Restricting plastic in this way meant countries such as Turkey took even more waste, with 40 per cent of plastic heading there in 2020.[6] At the time of writing Turkey had just announced restrictions on plastic imports, following reports of waste leakage into surrounding environments. These new restrictions have led to the question of which country will import it next – and can they actually take the volume of waste? The reality is that the UK needs to stop exporting this problem to countries that cannot manage it, but have low labour costs. Companies need to pay more for proper sorting and waste management. In the future we are likely to see more export restrictions as the government looks to improve the transparency of the recycling system and reduce waste leakage.

While National Sword led to some very extreme market behaviours in the short term, there are also a lot of positives that have come out of the policy, which mandates that waste should be sorted in its country of origin and managed effectively. Countries that had assumed someone would always want their low-quality waste are now ensuring they are reducing contamination by efficiently collecting and sorting, thereby increasing the quality of waste.

In the long term, National Sword will be seen as a pivotal moment that transformed global attitudes to waste. Ownership and responsibility instantly passed back to the country that created the waste – no longer could it simply be considered another country's problem.

What has the carrier bag charge achieved?

From 5 October 2015 in the UK, all single-use carrier bags legally had a charge applied of five pence per bag, which increased to ten pence in May 2021. The change in price also removed the threshold that meant the charge only applied to bigger retailers. The government reported a drop of 83 per cent in carrier bag usage between 2014 and 2016, but is this true?[7]

Bags for strife

The carrier bag charge has certainly had an impact: the charge levied on plastic bags went to charities, totalling annual donations of around £100 million, although this will reduce each year as usage decreases.

A single-use carrier bag is defined as a thin (70 microns or less) bag with handles that has not been used before. This means thicker 'bags for life' and things like woven bags are not included in the definition. There are also some exemptions relating to bags for goods like medicine and loose food products.

Customers are now far more likely to bring an old bag into a store when shopping. However, the exemptions to the definition of 'single-use carrier bag' mean the headlines do not match reality. In many stores now you can only buy bags for life or the equivalent, so thicker plastic is now the norm. This means the 83 per cent decrease in carrier bag usage is misleading, as it does not take into account the number of exempted bags sold, the quantities of which are not reported in the same way. A study from Greenpeace showed an average household purchased fifty-seven 'bags for life' in 2020,[8] by weight a significant increase in plastic usage from their single-use counterparts, although reporting on the charge would have you believe otherwise. Co-op actually reported that this shift to bags for life amongst supermarkets has led to an increase in plastic use of 440 per cent.[9]

The Environment Agency assessed how many times an alternative carrier bag needs to be reused to match the environmental impact of a single-use plastic bag: paper bags needed to be used three times, bags for life four times and a cotton bag a whopping 131 times.[10] This is a great example of where the media and reality present a different picture. Retailers are lauded for switching to paper bags, which generally survive one trip, whereas their 'single-use' plastic counterparts can in fact be used several times on average before breaking. If, for example, a single-use bag could survive five trips to the store, it would take nearly thirteen years of weekly shops for the cotton bag to have an equivalent environmental impact.

The commonly cited difference between single-use carrier bags and their alternatives is the image of plastic bags ending up in the ocean. What happens to carrier bags after usage does need to be considered when assessing the environmental impact of the options available. However, in a perfect world where waste is collected and recycled locally, this is a useful example of where plastic has a place. We should be investing in collection and recycling, tracking our waste and ensuring it is well managed, rather than simply (and perhaps unwittingly) shifting our environmental impact to somewhere else in the supply chain.

So next time you see a statistic about how amazing the carrier bag charge has been in reducing usage of carrier bags, an argument that also advanced the commitment to introduce deposit return schemes (page 56), remember it is not the complete picture. The levy is likely to have adjusted behaviour, but we cannot know whether the environmental impact has been equally positive.

What are microplastics?

Microplastics are defined as any plastic less than 5 mm in size, which can include small pieces of plastic that have broken off products, right down to microscopic  particles invisible to the human eye. Microplastics are everywhere! An estimated 270,000 tonnes of them come from tyres alone per year.[11] Microplastics that are easy to eradicate, such as microbeads, have already been banned. However, there are many more generated from clothing, larger plastic products breaking down and dust from cities.

Fragmented thoughts

Microplastics are a relatively new phenomenon, due to the rapid increase in our use of plastics, and they can come from many sources. Not much is understood about the potential health implications of microplastics because of their fairly recent emergence. You will find the plastics industry demonstrating evidence that microplastics are harmless, whereas environmental organisations will put forward examples of the harm they cause; the reality is that nobody really knows yet.

What we do know is that they are everywhere; they can be found in drinking water, the air we breathe and our oceans. Every time you open a plastic bottle, tiny plastic fragments will break off; a clothes wash will release an estimated 700,000 microfibres;[12] and every time you drive your car, tiny particles of plastic are worn away due to the friction of the road on your tyres. These plastics follow air flows or water paths, tending to settle in our oceans, where they are ingested by wildlife.

Most microplastics come from non-packaging applications: tyres and clothing are two of the main contributors, making up around 50 per cent of microplastics found in the ocean.[13] To reduce microplastics in your clothes washing, you can buy an external filter or a washbag which traps some of the microplastics. Amazingly, just your washing settings can make a difference: quicker washes which use less water reduce the quantity of microplastics generated, as does hang-drying rather than tumble-drying.

What has been banned?

In 2020, bans were introduced on plastic straws, stirrers and cotton buds, following the removal of microbeads from cosmetic products. These are not products that have been banned altogether, just their plastic varieties.

Stirring things up

Unnecessary packaging and products should be removed; there is little debate about that. Straws, stirrers and cotton buds do not need to be made of plastic, and therefore these bans seem fitting. However, all that happens when a material is banned is a move towards using other materials in its place. Paper straws are one example: since they are not made entirely of plastic, they are exempt from the ban. However, as they break up when in liquid, they are poor replacements and often paper straws will contain some plastic anyway, rendering them unrecyclable as a composite material. Their creation potentially harms the environment just as much as a plastic straw, since on average, paper has a higher carbon footprint than plastic.

Straws are generally not vital (with a few medical exceptions). Where this is the case, we should ban the product, rather than the material.

In the case of plastics, they are not all created equal. It is likely that over time we will see voluntary commitments to remove polystyrene from packaging, as there are alternatives that function just as well and are recyclable. While this could lead to a ban, the government is hopeful that future legislation, particularly extended producer responsibility (page 216), could remove the need to introduce bans by applying a high cost to unrecyclable packaging.

In Europe, oxo-degradable plastic (page 224) is banned, due to its tendency to break down into microplastics faster than conventional plastic.

Where do ocean plastics come from?

Aside from microplastics blowing and washing into the ocean, where do ocean plastics come from? Discarded commercial fishing materials make up a large portion of the waste. There are also some key rivers that carry a lot of waste to the ocean, which is then transported around the world by oceanic currents.

River deep, waste mountain high

When we hear the term 'ocean plastic', we may have an image in our head of waste being dumped from the side of a ship. In reality, this is a tiny percentage of the waste which reaches the ocean; poor waste management and littering are actually the key culprits.

There are 192 countries with a coastline and, according to an organisation called The Ocean Cleanup, approximately 80 per cent of all ocean plastics are thought to come from over 1,000 rivers in the world; this number was reported to be significantly lower, at around ten rivers, even as recently as 2017.[14] However, in recent studies the understanding of how rivers which pass through populated areas can be affected by waste has improved and broadened the scope of forecasts.[15]

Major countries which contribute to ocean plastic import a lot of waste from other countries, so we cannot wash our hands of the problem. The UK exports waste to countries where it could eventually leech into river systems and make its way to the sea, where it is transported by five major gyres, or large rotating ocean currents, into the 'great garbage patches', as they are known (which are anything but great).

The waste we export is likely to be paper and plastic, as a lot of our metals and glass are recycled in the UK due to their value. Recycling our own waste locally and reducing the exporting of waste to countries that are becoming overwhelmed with waste would have the single biggest impact on the amount of ocean plastic originating from the UK.

The majority of larger plastic waste is actually discarded commercial fishing gear, with estimates suggesting over 70 per cent of large plastic in the ocean is fishing-related.[16]

What sort of waste is in the ocean?

The waste that can be found in the ocean comprises microplastics, rubbish from bad waste management and  discarded fishing equipment. The waste can be ingested by wildlife or caught in gyres which circulate to form the 'great garbage patches'.

Beached fail

Normally, waste will tend to break down into smaller and smaller pieces, but it will not disappear. Therefore, the waste in our oceans accumulates and the build-up worsens as our consumption increases.

Environmental research consultancy Eunomia looked into where ocean plastics come from and found over three-quarters comes from land-based activities, with the remainder coming from beaches (5 per cent) and fishing (15 per cent). Around 95 per cent of waste that reaches the sea will then sink to the ocean floor.[17] With regard to microplastics, the majority is produced by tyres, followed by pellet spills (where plastic from recycling companies accidentally ends up in the ocean) and finally textiles, when tiny fibres from synthetic fabrics come loose when we wash our clothes.

It is not just plastic that causes problems in our oceans, although it does make up around 60 to 80 per cent of all ocean debris.[18] Other materials which can be found include metal, glass, paper and cardboard. All poorly managed waste has the potential to end up in the ocean, highlighting the need to improve collection and recycling practices.

Wildlife can mistake debris for food, causing them to choke or for waste to block their digestive systems. Larger items such as fishing nets can cause strangulation. Ingesting microplastics could cause chemicals to be absorbed, weakening an organism's immune system and making them more susceptible to diseases and other infections.

Is packaging use decreasing?

Images of oceans filled with plastic and affected wildlife appear to have changed public views and the choices they make when buying products. So, has the volume of packaging sold actually decreased in line with public opinion?

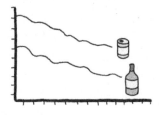

Flexible numbers

The Environment Agency publishes data each year on how much packaging is used by UK companies. This is calculated for the producer responsibility regulations (page 14) and is publicly available.

The latest weight data available at the time of writing is the 2020 data. When compared to 2019, most packaging types actually increased, which appears to be linked to Covid-19 and the change of packaging use during the year we were stuck at home.

The biggest increases are in glass and aluminium (up around 4 per cent and 9 per cent respectively), thought to be because we stopped visiting the pub, where drinks are served draught, and replaced this with bottles and cans at home.[19]

Paper and plastic also increased, albeit to a lesser degree. Cardboard boxes were delivered to houses more frequently, and replaced some of the transit packaging that businesses generate. Home cooking creates more packaging as we do not buy products in bulk as a restaurant would.

Covid-19 has certainly had a significant impact on the nation's packaging use, increasing its use and changing material composition as a result of remaining at home. Perhaps, then, it is fairer to compare 2019 with 2018, which was a time when plastic was in the media constantly and our lives were relatively normal.

Compared with 2018, in 2019 there were small increases (1–2 per cent) in glass and aluminium, and a 4 per cent drop in plastic use.

Normally, the data is quite steady; while sales tend to increase, most manufacturers actively reduce the weight of their packaging, which tends to balance the figures. A 4 per cent decrease in plastic was extremely significant and reflected a changing public attitude.

This thinking was supported by an increase in materials that could typically replace plastic, such as aluminium cans or glass bottles. There was an 8 per cent increase in a catch-all category which is not defined as any particular material, which can include non-traditional packaging like fabrics and cork. This figure suggests a big increase in plastic alternatives.

As we start coming out of lockdowns and return to a sense of normality, the question is, will consumers maintain the 2019 trend of seeking out alternatives to plastic, or have we got used to our 2020 lifestyle, with more home deliveries and cooking which increases packaging use? Future data will certainly give us a lot of insight into consumer sentiment.

Problems exist in the world of waste; the system is not perfect and never will be, and our drive for convenience may create unnecessary waste. However, academics, scientists and governments are working hard to make things better and innovation is rapidly changing the perception of waste, turning it more into a resource. Let's explore some of the best innovations.

10

The ~~Rubbish~~ Better Future

The world is changing, and the focus on sustainability, climate change and recycling will drive rapid innovation, some of which we are seeing already. This section contains an overview of some of the most significant and interesting developments, which are only a few years away.

What does future packaging legislation look like?

On 31 January 2020, the UK left the EU, heralding a new era of waste management. What has become clear, based on the announcements that followed, is that the UK is committed to being at the forefront of tackling environmental issues, and is likely to create and develop ambitious legislation around recycling and waste management.

With so much focus on environmental issues in the public domain, it is perhaps unsurprising that politically there is a focus on improving legislation.

There are lots of environmental law changes coming, but four main pieces of future legislation that affect the world of packaging include: deposit return schemes, consistency of collections, plastic packaging tax and extended producer responsibility, which is an improvement on the current producer responsibility system (page 14).

Deposit return schemes

Deposit return schemes add a small additional cost to beverage packaging, which is recovered by the customer when the packaging is returned to a designated point for recycling (page 56).

The current target for the introduction of deposit return schemes is 2024 in England, Wales and Northern Ireland, with Scotland aiming for 2022. The scheme is being designed at the moment with a focus on the value of the deposit, who should be responsible for managing the scheme and how to protect the system from fraud, which has affected schemes abroad. It is possible the launch will be delayed beyond 2024, as there are many key elements to be agreed before the legislation can be finalised.

Consistency of collections

This future legislation is focused on standardising the collection of waste across all local councils in England. Standardising collections will have a very positive impact on household collections, as it will increase confidence in the recycling process and reduce confusion about what householders can recycle.

It is likely the mandatory collections will be announced in 2023 and will include food waste and flexible plastic. However, it will take time to implement nationally as councils will have to wait until waste management contracts are up for renewal before they can bring their waste collections in line with the required national standard.

Plastic packaging tax

Announced in 2018, and coming into force from April 2022, this legislation will apply a £200 per tonne tax on any plastic packaging that contains less than 30 per cent recycled content.

This is an interesting government response to the lack of plastic recycling. As explained in this book, the main issue with plastic collection and recycling is economic viability: everything *can* be recycled, but the process of doing so must be scalable and sufficiently valuable for recycling to take place in reality. Placing a tax on plastic that does not contain recycled content immediately increases the value of recycling. Brands and manufacturers will agree long-term commitments with recyclers to avoid the charge. So by rapidly increasing the value of recycled plastic and by extension its recycling rate, this tax will have a big impact.

Extended producer responsibility

The current producer responsibility system covers around 10–20 per cent of waste management costs. Essentially, this legislation places a cost burden on brands and retailers. In future, under extended producer responsibility legislation, these companies will have to cover 100 per cent of the waste management costs.

Scheduled to come into force in 2024, the extended producer responsibility framework will calculate the total cost of household waste management, including all the collections, transport, recycling and litter management costs. This total cost will then be passed back to the manufacturers of the product.

The fee a manufacturer will pay is known as a 'modulated fee'. This means a company manufacturing a product which is difficult to recycle will pay more than a company placing an easily recycled product on the market. This incentivises manufacturers to use colours and packaging types that are widely recycled.

Together, these four pieces of legislation cover the collection of packaging (deposit return schemes and consistency), the economics (plastic tax) and the costs of household waste (extended producer responsibility). There is no doubt they will have a significant impact on the UK waste industry and future recycling rates.

How is chemical recycling expected to develop?

Chemical recycling (page 88) is likely to be a big part of future waste infrastructure, as it can be used to process a wider range of materials and is less sensitive to food waste contamination than its mechanical alternative.

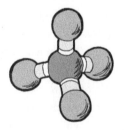

A chemical breakdown
Chemical recycling is not new; however, it is currently the focus of a lot of investment (and hope) as companies try to solve the plastic problem, particularly with hard-to-recycle packaging,

such as flexible plastics. Think of it like 'un-baking' a cake: where mechanical recycling would smash the cake up, chemical recycling aims to return the cake to its original ingredients.

A number of large chemical recycling plants are now planned for the UK and the recyclers are actively signing agreements to arrange for waste to be brought to the plants and planning where to send the oil output produced by the pyrolysis process. By 2025, it is likely that tens of thousands of tonnes of material will be chemically recycled.

Economically, chemical recycling is quite an expensive process, and therefore it will be developed in line with future legislation (page 216) which will help support the costs involved, particularly the plastic tax, which will cause brands to invest in chemical recycling, as there is no other way to recycle certain types of plastic. For example, to recycle a chocolate wrapper into a new film, it must go through a chemical recycling process.

The downside to chemical recycling is that it generally requires a specific set of materials to create high-quality output. Too much of the wrong kind of plastic, and the oil produced will only be of use as fuel rather than as a building block for new plastic. For example, pyrolysis works best with an absence of oxygen, so if a load contained a lot of PET plastic, which contains oxygen, it would reduce the quality of the output. This means that, while chemical recycling has been hailed as a miracle process, real-world applications of the output generated by the recycling facilities could be limited, based on the sources of packaging.

What are bioplastics?

The rapid growth in plastic as a material
was largely driven by the fact that plastic
could be made from the by-products of
abundant fossil fuels. However, plastics
can also be made from more sustainable
and renewable bio-based materials, such as
plants, starch and natural oils. Bio-based plastics behave exactly
like their petroleum equivalent and the term refers purely to the
source, not what happens to the product at the end of its life.

Recycling plant

The very first types of plastic developed were bioplastics:
Parkesine, the first plastic, developed in 1856, was plant-based
and used cellulose (page 36). In contrast, it was not until 1907
that the first fully synthetic plastic, containing nothing found in
nature, was developed. Bioplastics are now a very small part of
the overall plastic market (less than 1 per cent) and they are
likely to be found in disposable cutlery and cups when buying
food and drink on the go.

The inclusion of 'bio' in the name causes confusion, as there
is an assumption that this means it is biodegradable (page 222)
or compostable (page 225). Like fossil fuel-based plastic,
bioplastic may break down, but it is equally likely to behave like
a traditional plastic and not biodegrade easily or be classed as
compostable. Remember, the 'bio' in its name refers only to the
source of the plastic, not how to dispose of it.

There are two types of bioplastic: those that are chemically identical to their fossil fuel counterpart, and those made with bio-based chemicals, which do not have a counterpart made of fossil fuel. This means there is a bio-PET, bio-PE and bio-PP, which behave and perform in the same way as their counterparts derived from fossil fuels. They are therefore recyclable in the same way, but will not biodegrade quickly. Dedicated bioplastics made with bio-based chemicals may have been designed to degrade.

Their name implies they are more sustainable to produce, and certainly they are lower in toxins and use renewable materials, all of which is good. Compared to traditional plastic, they also reduce carbon emissions, as their growth removes the carbon dioxide from the atmosphere that they emit as they decompose at the end of life.

However, as usual, lifecycle analysis suggests things are not quite as simple and in lots of ways bioplastics may not be as sustainable. Fertilisers and pesticides are used to grow the crops to make bioplastics and the chemical process needed to turn them into plastic is significant.

From an entire lifecycle perspective, PP, HDPE and LDPE were all deemed to be better than their bioplastic equivalent in a study completed by the University of Pittsburgh.[1]

What is biodegradable plastic?

First things first: *everything* is biodegradable, as eventually everything will break down; it just might take a very long time. For this reason, the phrase

'biodegradable plastic' can be misleading, and accusations of 'greenwashing' are common.

Biodegradable plastic is seen as a potential replacement for packaging which is currently difficult to recycle. The term 'biodegradable plastics' should mean plastics that will decompose over a defined time, and under certain conditions.

A degrading solution

The government is trying to standardise the term 'biodegradable' to avoid confusion and greenwashing. In October 2020, it announced a new set of standards to govern the term. To meet the standards, biodegradable plastics must break down to a wax which does not contain microplastics within two years.

Currently, most biodegradable plastics require heat to break down, which is not readily available in the normal environment or the ocean. Often, consumers interpret 'biodegradable' to mean that litter will degrade if left in the natural environment, which is not the case due to the lack of heat.

The major issue with true biodegradable plastic is that it cannot go in the recycling bin! It looks and behaves like a conventional plastic but cannot be recycled in the usual way, due to its potential to weaken in the future. This means it must be put in the normal rubbish bin to be sent for incineration or landfill, where it is unlikely to decompose. If it is placed in the food waste bin, which feels like the right thing to do, it may not be identified as a biodegradable plastic and therefore is likely to be removed and sent for incineration or landfill anyway.

What is oxo-degradable plastic?

Oxo-degradable plastic is a specific type of plastic which is neither a bioplastic nor biodegradable. To create an oxo-degradable plastic, a small amount of metal is added during the production of normal plastic. The metal causes the plastic to break down faster if it gets into the environment, mimicking biodegradability. It is fair to say this type of plastic has shown the divisions that exist across countries in their approaches to waste management.

Heavy metals

The fact that oxo-degradable plastic is made of normal plastic means that it can be recycled in the usual way, unlike biodegradable or compostable plastic. Oxo-degradable plastic will also mostly break down over two years, but it will break down into microplastics (page 205), rather than at the molecular level, like biodegradable plastics. This means oxo-degradable plastics tend to be worse for the environment than the problem they are attempting to solve.

This makes them a confusing plastic to the average consumer. The logic is sound: a plastic that can be recycled but will break down if it is littered appears to be very positive. But this message could lead to a significant increase in litter, as it suggests littering a plastic that will break down is environmentally friendly (it is not!). In addition, the fact that oxo-degradable plastic breaks down into microplastics is a huge issue, as this causes environmental damage. It is better for the plastic to remain in large pieces as in this state there is a chance it will be captured and removed from the environment.

For this reason, the EU introduced a ban on the use of oxo-degradable plastics, which came into force in 2021. Following Brexit, the UK is not required to follow this ban, but it is likely the government will nonetheless do so.

In stark contrast, Saudi Arabia has mandated that all PE and PP packaging must contain an oxo-degradable additive, effectively mandating its use. This is a perfect example of consumer feeling trumping the science.

What is compostable plastic?

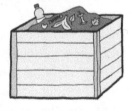

Compostable plastic is slightly different from biodegradable plastic, as it will break down over a certain period of time – most likely in an industrial composting facility, though certain plastics can even be suitable for home composting. There is a standard (EN13432) that compostable plastic must meet in order to carry the symbol for composting (page 131). To meet the standard, compostable plastic must not leave any toxic residue, unlike biodegradable plastic.

Down to earth

'Compostable' suggests that you could put the product in a compost bin in your garden and it will break down. This is not normally the case unless it has the symbol for home composting (page 132).

Compostable plastic will only break down under certain conditions, usually requiring air, moisture and sunlight. It will not break down in landfill, as it will be deprived of these key elements.

The EN13432 standard requires 90 per cent of the plastic to break down into carbon dioxide, water and biomass over a period of twelve weeks; however, this is in industrial conditions, which is a controlled environment. Therefore, the main issue with compostable plastic is that the name suggests you can leave it lying around outside and it will eventually just disappear. Actually, it needs to be collected and sent to special facilities.

The complexity of collection is one of the biggest arguments against compostable packaging. Compostable plastic cannot be recycled in the traditional way. However, it is perfect for lining a food waste caddy, since if it is left in the process, it will break down, as it has entered the industrial process with the food. Unfortunately, most industrial composters accepting food do not want plastic in their process, so at the moment compostable plastics should go in the normal waste bin.

In order for compostable plastic recycling to be effective in the future, more recyclers will need to accept it, and markings on compostable plastic will need to be clear and visible. This will ensure they meet the required standard and are labelled accordingly, which is the best way to make sure they are collected and recycled.

For all these reasons compostable plastic is perfect for closed environments such as festivals, as it can be collected and sent to industrial composters with no contamination. It is also useful as a material for things like carrier bags, where it can be used as caddy liners.

What is mono-material packaging?

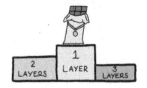

Modern packaging is often designed to look good rather than to be easily recycled. This means there are lots of different materials used in packaging. However, there is now a move away from composite materials and towards 'mono-material', where an entire piece of packaging is made of just one type of material.

Back to basics

Composite packaging (page 107) can serve a purpose. For example, the lid of a plastic bottle is sometimes made of a different type of plastic from the bottle so that it is strong enough to stay on, or a chocolate bar wrapper will need a mix of plastic and aluminium to ensure protection from external conditions.

The major shift we will see in packaging over the coming years will be a trend towards simpler, single-material packaging. New innovations in packaging materials and recycling technology will allow for packaging to still have a protective barrier, despite being made of one material. This will make it much easier to recycle and increase our confidence in what we can put in the recycling bin: the more packaging is made from one material, the easier it will be for consumers to quickly identify and correctly recycle it.

A good example of the shift to mono-material is the recent developments in the pouch market. Pouches are a lightweight and very effective packaging material, but as they are made of a combination of plastic and aluminium, they are hard to recycle. Pouch manufacturers have created versions that are made only of polypropylene (PP) for use with pet food, which will be far easier to recycle.

Large brands are keen to ensure their products are made with a single material. As a result, we are likely to see an increasing number of interesting developments in this space that will pave the way for increased recycling.

What does 'pay as you throw' mean?

'Pay as you throw' is a system of charging for the amount of waste collected from the household, normally focused on waste destined for landfill. In the UK, collection and recycling costs are currently covered by council tax. This alternative charging system tends to ensure people recycle as much as possible, as they will pay more if they do not.

Weighing up the costs

In Europe, Austria was the first country to use 'pay as you throw' in 1945. Electronic tracking became key for 'pay as you throw' to work effectively, so it was not until the 1980s that it became popular. The first country to use electronic tracking and billing was Germany. The system typically leads to a reduction in normal rubbish and an increase in recycling.

'Pay as you throw' relies on consumers purchasing bags or tags to ensure waste is taken away. In the UK, recycling costs are covered by local authorities and charged through the tax system. This means everyone pays the same, regardless of how much waste they throw away, and so there is no financial incentive to reduce waste or increase recycling.

One of the major concerns of a 'pay as you throw' system is that it could also increase fly-tipping by householders seeking to avoid charges.

'Pay as you throw' works well to help people think differently and incentivises recycling. This is a model the UK has considered and potentially could use in the future to increase recycling and reduce waste. However, it does not tie in well with future legislation changes, where the cost is likely to sit with the brand owner (page 216).

How will the frequency of collections change?

One of the most effective and proven ways to increase recycling and reduce the volume of normal waste is to increase the frequency of recycling and food collections from households and reduce the number of general waste collections accordingly.

Time for change

Over three-quarters of local authorities collect general waste fortnightly. If this is tied in with frequent recycling collections it is an effective way to increase recycling rates. It is likely that in a few years, there will be a number of local authorities moving to collect waste going to landfill only once a month.

Some councils alternate between general waste and recycling collections, while others collect recycling weekly. There is clear evidence that a weekly recycling collection combined with a less frequent general waste collection increases recycling rates. However, inevitably the choice of each council comes down to cost. This could change with the introduction of extended producer responsibility and consistent collection legislation (page 216), which is likely to fund and promote best practice.

Less frequent general waste collections are already being trialled in a few areas and have been met with concern about an increase in pests, such as rats and seagulls, due to waste being left outside for a longer period between collections. This can be a concern, particularly over a hot summer, in houses of larger families or those with babies in nappies who will naturally produce more waste.

Conwy County Borough Council in Wales reported that in the first year of trialling monthly collections there was a 14 per cent increase in recycling and a 31 per cent decrease in normal waste.[2]

What can technology do for the waste industry?

The waste industry relies on technology to improve collection, sorting and recycling at large facilities, which are expensive to run. However, due to long-term contracts with local authorities,

and a resistance to change, this technological revolution can be more gradual than in other industries. So what might the waste industry look like in ten years?

Collection

Our kerbside collections will adapt to remove beverage packaging that will instead be collected in deposit return schemes (page 56). The new deposit return scheme will be technologically driven, dramatically reducing contamination in PET. For business collections, sensors are likely to be added to bins which will tell collection companies when they need emptying, leading to the waste collection process becoming almost completely automated.

A significant amount of data is likely to be gathered around collection in the future, as knowing what is being discarded and where is going to be essential to the management of future legislation. This will help the UK move to a truly circular economy.

Sorting

Improving the quality of recyclate will be important to increase future recycling rates. This means facilities will upgrade recycling equipment to reduce contamination, with more advanced cameras, tagging (page 235), separation and artificial intelligence to distinguish by packaging type as well as chemical composition. This technology will also be able to distinguish specific brands, so companies producing these brands will be able to determine their individual recycling rates, rather than overall material rates.

Recycling

Chemical recycling (page 88) will grow significantly and allow the recycling of almost everything that cannot currently be mechanically recycled. This will need to take place in line with sorting improvements, to ensure low contamination levels. State-of-the-art mechanical recycling facilities will be built to deal with a wider range of materials, increasing recycling rates.

What is the future of MRFs?

MRFs (page 66) have typically been built over a number of years and added to as new technology becomes available. This modular and unpredictable expansion has resulted in inefficient MRFs that 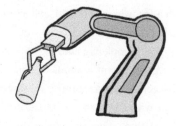 would be built differently if starting from scratch today. This has led some local authorities to join together to build 'super MRFs', facilities designed with the latest technology, to maximise throughput and a broader range of material input and output.

A material world

MRFs have a problem: they are designed to process a specific set of materials, based largely on the waste composition of the average household. This will change as the introduction of new legislation, like deposit return schemes (page 56), will remove bottles and cans from mixed collections, resulting in high-value material suddenly disappearing from a MRF's input. Operators know this and are already considering how they change their plants in the future to accept a wider range of materials. If they do not act now, these ageing plants may soon become economically unviable.

The plastic packaging tax (page 216) will also have an impact. Manufacturers will need high-quality material to use in food grade applications. This will entail both an improved sorting process and additional costs, which will be met with the introduction of extended producer responsibility.

Investments are currently being made, with MRFs seeking more accurate and efficient machinery. For example, robotic arms that grab waste using artificial intelligence are replacing air flows, and tagging packaging at the point it is made will become the norm (page 235). Every year our packaging increases, and plants built ten years ago need to evolve to cope with ever-increasing and evolving waste.

How could we recycle the hard-to-recycle plastics?

Already, there are specialist collection routes (page 52) that enable the collection of crisp packets, biscuit and chocolate wrappers and other hard-to-recycle plastics, albeit in relatively  low volumes compared to the total number manufactured.

Where are these plastics going if chemical recycling is not yet manageable at scale?

Board meetings

All plastics can be made into new products; they just may not be turned back into plastic packaging. For example, hard-to-recycle plastics are currently being turned into boards that look a bit like wood and can be used as an alternative to plywood in the furniture and construction industry. The UK imports around 3.3 million cubic metres of wood panelling a year,[3] so using plastic instead could improve deforestation rates in other countries and reduce the carbon footprint of importing this wood.

There are a number of ways of making board out of plastic, which all involve melting the plastic first and forming it into a board shape, either using vacuum moulding or extrusion. The board has a plastic skin glued on either side to increase strength and ensure it is smooth. These boards can be chopped into shape and are typically used to make benches, boarding for retail stores or even new bins.

One of the major advantages of creating board out of plastic, rather than using wood, is its weather resistance. The farming industry has been particularly supportive of the process, using recycled plastic for fencing which is exposed to rain. The board may be slightly more expensive than its plywood equivalent, but the fact that it can survive multiple seasons makes it significantly cheaper over its lifetime.

What is packaging tagging?

Packaging tagging, using chemical tracers or digital watermarks, will become a key part of sorting at MRFs (page 66) and recycling plants in the future. Tagging allows MRFs to identify both the chemical composition and the actual application of the packaging. For example, a PET tray will be identifiable as a PET tray, rather than just PET.

Tracing ~~paper~~ plastic

The sorting of plastic, currently achieved with NIR (page 71), could change with the introduction of digital watermarks, which would enable better sorting and reporting of plastic waste. This would also constitute a move to a circular economy (page 116) as a specific piece of packaging could be tracked from manufacture to recycling.

There are two ways of 'tagging' packaging: using chemical tracers or digital watermarks. This essentially means the packaging will be covered with a unique identifier that will be invisible to the naked eye. However, once the packaging reaches the recycler, this code can be read by machines and used to identify the material the packaging is made from and how to sort it. The watermark would cover the packaging repeatedly so that it could be read even if the packaging has been scrunched up.

This technology has been tested and piloted through a collaboration called the HolyGrail project, which has united twenty-nine partners since 2016. The project aims to create a 'barcode of recycling'. The project has concluded an initial three-year trial to test the new technologies that exist and understand how this could be taken forward at scale. While it is providing interesting results, a lot of work will still be required to understand how this method of sorting can be scaled up across the industry.

What if we did not recycle?

What would happen if we did not bother with recycling? What would happen if we only used everything once?

One chance

The first thing that would be required is a lot more landfills (page 195) or increased construction of energy from waste plants (page 197). In the UK, we would need around three times more landfills, which would take up more land and create more dangerous gases, such as methane and carbon dioxide, which are released when waste breaks down. These gases speed up global warming. If we focused on energy from waste, this would create an energy supply that could potentially be more carbon-intensive than our future low-carbon supply.

Natural resources would also run out more quickly. More trees would be cut down for paper, cardboard and wood. Oil would be used up faster to meet the requirement of our increasing raw plastic use. Mining for metals would rise. The energy requirements of creating everything from new would significantly increase global warming. All these resources will run out eventually and failing to recycle would make this happen faster.

The inescapable fact is that recycling is the best way of protecting our planet, considering our current consumption habits and lifestyle. In order to remove packaging and reduce food waste, our lifestyles will need to change drastically, focusing on more local produce and less supermarket and online shopping. If we cannot all do this, the best things we can do for our planet are to reduce our packaging usage by avoiding pointless products and packaging, assessing the right material for the right product (which can include plastic) and eliminating things that cannot be recycled.

Conclusion

Everything can be recycled.

Everything!

If that is all you take from this book, I will be happy. As we have seen, most of the issues around recycling stem from the economics of certain materials, investments in technology and the kind of packaging we find on our supermarket shelves.

So, let us take everything we have learned and make the world better together. This entire book can be summarised in some very simple rules, outlined below.

Choose wisely

- Select lightweight and sensible packaging for the product you are buying. Toilet rolls individually wrapped in paper may not always be better than a multipack wrapped in plastic.
- Moving away from plastic should not be your only goal – in fact, in a lot of uses plastic is better than alternative packaging. Always think critically about the environmental impact of alternatives to plastic.
- Avoid difficult-to-recycle materials; buy non-PVC cling film, avoid polystyrene, oxo-degradable packaging and anything that does not have a clear and evident recycling route.

Recycle well

- Wash and dry materials before they are placed in the bin, reduce food contamination and cover paper and cardboard when outside.
- Do not try and recycle things at home that definitely should not be recycled, like nappies and batteries. Some products will create serious contamination issues and should be thrown away in general bins (nappies) or collected in dedicated bins (batteries).
- Not all plastic is the same – look for the number and check whether it can be recycled.
- Check locally – this book is general guidance but, for now, each local authority will collect and recycle differently, so please check your local council website to find out what can and cannot be collected in your area.

Be vocal

- If your council does not collect glass, or your local coffee shop will not take back your cup, or you notice anywhere else not doing something they should to facilitate good recycling practices, demand they do. Write to them, help them understand best practice and instigate change.
- Spread the word. The main reason I wrote this book is to try and distribute these tips and tricks beyond the waste industry. Don't let it stop with you – keep passing your favourite facts on and encourage people to find out what happens to their waste once it has gone into the bin.

Thank you for taking the time to read *The Rubbish Book*, a book that has been in my head for a number of years. I am and always will be extremely passionate about the recycling industry and how we can all play our small part in helping the world to become more sustainable.

I hope this book has proved useful in your travels towards becoming a sustainable consumer, and I wish you the best of luck on your journey.

Resources

The world of waste is rapidly evolving, and there's a lot of information out there. Below are some of the websites that I find most useful when trying to stay up to date.

360 Environmental – https://www.360environmental.co.uk/
An independent team of waste management experts that specialise in helping companies to understand and comply with a wide range of waste legislation compliance issues.

Alupro – https://alupro.org.uk/
Alupro is an industry-funded, not-for-profit organisation with over thirty years of experience representing the UK's aluminium packaging industry. Alupro runs the programmes Metal Matters and Every Can Counts.

BBIA – https://bbia.org.uk/
BBIA is the UK trade body for companies producing bio-based and biodegradable products and promotes the circular bioeconomy.

British Plastics Federation – https://www.bpf.co.uk/
The British Plastics Federation (BPF) is the world's longest-running plastics trade association. It was established in 1933 and has represented and promoted the UK plastics industry ever since.

Confederation of Paper Industries – https://www.paper.org.uk/
The Confederation of Paper Industries (CPI) is the leading trade association representing the UK's paper-based industries.

Ecosurety – https://ecosurety.com/
Ecosurety is one of the UK's leading packaging, WEEE (Waste Electrical and Electronic Equipment) and batteries compliance schemes, working with companies under producer responsibility legislation (page 14) and over the last twelve years as I moved from graduate to CEO, the company that taught me most of what I know.

Ellen Macarthur Foundation – https://www.ellenmacarthur foundation.org/
The Ellen MacArthur Foundation develops and promotes the idea of a circular economy. They work with business, academia, policymakers and institutions to mobilise systems solutions at scale, globally.

ESA – http://www.esauk.org/
The Environmental Services Association (ESA) is the trade body representing the UK's resource and waste management industry.

Eunomia – https://www.eunomia.co.uk/
Eunomia is an independent consultancy dedicated to helping their clients to achieve better environmental and commercial outcomes.

Flexible plastic fund – https://flexibleplasticfund.org.uk
A project I worked on for two years, which launched in 2021. This fund looks to increase the value of flexible plastics. Brands pay into a fund which pays out when recycling is proven.

Hubbub – https://www.hubbub.org.uk/
Hubbub designs fun and engaging consumer-facing campaigns to inspire ways of living that are good for the environment.

Incpen – https://incpen.org/
Incpen stands for 'Industry Council for Packaging and the Environment' and is a membership organisation working with governments and businesses.

Keep Britain Tidy – https://www.keepbritaintidy.org/
Keep Britain Tidy is an independent charity with three goals – to eliminate litter, end waste and improve places.

LetsRecycle – https://www.letsrecycle.com/
LetsRecycle is a leading independent website for businesses, local authorities and community groups involved in recycling and waste management.

Love Food Hate Waste – https://www.lovefoodhatewaste.com/
Run by WRAP (see below), Love Food Hate Waste aims to raise awareness of the need to reduce food waste and help us take action.

National Cup recycling scheme – https://www.cuprecyclingscheme.co.uk

A fund to provide value to used beverage cups, increasing the value of each tonne collected. It is supported by major brands who also have to collect back cups as part of their commitment.

OPRL – https://www.oprl.org.uk/

The OPRL scheme aims to deliver a simple, consistent and UK-wide recycling message on retailer and brand packaging (page 133).

Podback – https://podback.org

A project I worked on for nearly three years, Podback is a national scheme to ensure used beverage pods are collected and recycled. Providing dedicated collections to consumers, with a focus on increasing kerbside infrastructure.

Recoup – https://www.recoup.org/

Recoup aims to lead and inform the continued development of plastics recycling and resource management, be an independent voice of reason and educate the public and businesses on the recycling of plastics to protect the environment.

Recycle Now – https://www.recyclenow.com/

Recycle Now is the national recycling campaign for England, managed by WRAP (see below). Recycle Now helps people to recycle more things, more often.

Resource Futures – https://www.resourcefutures.co.uk/

Resource Futures is an environmental consultancy that works

with clients across the private, public and non-profit sectors to enable the positive management of material resources.

Terracycle – https://www.terracycle.com/
TerraCycle is an innovative collection company, focusing on specialist collections (page 52) of hard-to-recycle waste.

The Ocean Cleanup – https://theoceancleanup.com/
The Ocean Cleanup is a non-profit organisation developing advanced technologies to remove plastic in the oceans.

WRAP – https://wrap.org.uk/
WRAP (Waste and Resources Action Programme) promotes and encourages sustainable resource use through product design, waste minimisation, reuse, recycling and reprocessing of waste materials.

Glossary

Anaerobic digestion – the process where bacteria break down organic matter, without oxygen.

Aspirational recycling – also known as 'wishcycling': a consumer putting something in the recycling bin that cannot be recycled.

Authorised treatment facility (ATF) – permitted site which carries out the treatment of waste electricals.

Automotive battery – a rechargeable battery used to start a motor vehicle.

Bale – a load of waste compressed into a cube shape.

Bauxite – a rock comprised of aluminium oxide, which is the world's primary source of aluminium.

Biodegradable plastic – a type of plastic that will decompose over a defined time, and under certain conditions.

Bioplastic – plastics made from renewable bio-based materials, such as plants, starch and natural oils.

Black plastic – plastic dyed with 'carbon black' (see below), making it hard for near infrared to detect.

Bottle bank – a bin specifically designed for glass bottles, commonly found in supermarket car parks.

Carbon black – a carbon powder used as a pigment, made by burning hydrocarbons in insufficient air.

Cellulose – the main molecule in the walls of plant cells, helping to give them strength.

Chemical recycling – the process of recycling by chemically altering plastic back to its monomer and putting it back together as a recycled polymer.

Circular economy – an economic system aimed at eliminating waste by bringing it back into the system as a product, through recycling or reuse.

Civic amenity site – a facility where the public can bring household waste, also known as a household waste recycling centre.

Co-mingled collection – the process of collecting waste mixed, which leads to waste needing to go to a MRF (see below).

Composite – packaging that is made up of several materials, which are difficult to separate.

Compostable – materials that will break down completely into non-toxic components (water, carbon dioxide and biomass), given the right conditions.

Conference of parties (COP) – the supreme governing body of an international convention, composed of representatives of all parties.

Contaminant – a substance that makes something unsuitable through unclean contact.

Contamination – the act of making something unsuitable by contact with something unclean.

Corrugated boxes – cardboard boxes made of three layers, two smooth outer layers and a pleated inner layer.

Cracking plastic – the process used in 'chemical recycling' (see above) to convert complex polymers into simpler molecules.

Crude oil – a naturally occurring black liquid found in geological formations beneath the Earth's surface, also known as petroleum and oil.

Cullet – waste or broken glass that is in smaller pieces ready for recycling.

Deforestation – the act of cutting down a wide area of trees.

Density separation – using the varying density of different materials to separate them in liquid.

Depolymerisation recycling – one of the ways to 'chemically recycle' plastic, by breaking polymers into monomers, before feeding them back into production.

Deposit return scheme – a system where a deposit is paid at the point of purchase and returned when the packaging is given back for recycling.

Die – a hole, usually cut into metal, which when melted plastic is passed through will mould and shape it.

Digester tank – storage equipment which allows for 'anaerobic digestion' to take place.

Domestic recycling – recycling occurring within a local system, for example, waste generated in the UK being recycled within the UK.

Duty of care – a piece of legislation that requires the safe management of waste to protect human health and the environment.

Eddy current separator – a machine that uses a powerful magnetic field to separate aluminium from other waste.

Electrodes – a conductor of electricity used to make contact with a non-metallic part of an electrical circuit.

Electrolysis – a process that uses electrical current to drive an unnatural chemical reaction.

EU Waste Framework Directive – a legal act setting out a goal for EU members, in this case with the aim of helping member states protect the environment.

Expanded polystyrene – typically white and foam-like plastic used in packaging, usually surrounding fragile items like televisions.

Extrusion – the process of pushing melted plastic through a 'die' (see above), to create a plastic with a new shape.

Feedstock recycling – another term for 'chemical recycling' (see above).

Flake – a chopped or shredded piece of plastic, the first stage of plastic recycling.

Flexible film/plastic – plastic packaging that is able to be screwed up into a ball, like cling film, a crisp packet or a chocolate wrapper.

Fly-tipping – the act of throwing waste into an unauthorised location such as a roadside.

Food contact – packaging that is clean and sterile enough to be placed against food before sale.

Gasification – the process of heating plastic waste with air or steam to produce a synthetic gas, sometime referred to as syngas.

Glass aggregate – crushed glass (cullet – see above) which can be used in concrete.

Glass remelt – glass which is melted back into a usable product, such as bottles or jars.

Granulator – a device that chops plastic up until it passes through holes in a screen.

Green Dot – a symbol found on packaging, consisting of two intertwined arrows: it signifies a recycling financial contribution has been made, but is not used in the UK.

Greenwashing – conveying false or misleading information about a product or service which consumers may believe makes the company more environmental than they are.

Hard-to-recycle plastic – plastic that is currently not widely collected or recycled, for example, flexible film (see above), polystyrene or PVC.

HDPE – high-density polyethylene, a thermoplastic produced from ethylene.

Home compostable – a type of compostable plastic that will break down in home composting conditions.

Hydrocarbon – a compound made from hydrogen and carbon.

Industrial battery – a battery which is designed exclusively for professional use.

Kerbside sort – the act of sorting waste by waste collection crew at the point of collection.

Landfill – a site for disposing waste by burying it in the ground.

LDPE – low-density polyethylene, a thermoplastic produced from ethylene, mostly found in flexible form.

Lightweight – the act of redesigning packaging to weigh less.

Linear economy – the opposite of 'circular economy' (see above), where waste is disposed and lost after a single use.

Local Authority – an administrative body within local government.

Low carbon – power sources that minimise the emission of greenhouse gases.

Magnetic head pulley – a magnet that sits within a conveyor belt to allow magnetic material (like steel) to be separated from other material.

Materials Recovery Facility (MRF) – a facility where materials are sorted by type in order to be sold.

Mechanical recycling – the alternative to 'chemical recycling' (see above): the recycling of an item without significantly altering its chemistry, for example, using shredders or heat to melt it.

Metallised plastic film – flexible plastic film with a thin layer of metal, used for things like crisp packets and chocolate wrappers.

Microbeads – small pieces of plastic, usually less than 5 mm in size.

Mobius Loop – a symbol that indicates that the packaging is made from a material that can be recycled.

Mono-material – packaging made with one type of material.

Monomer – a molecule that can be bonded to other identical molecules to form a polymer.

Multi-stream collection – a collection where materials are separated either before or during collection, usually through having multiple bins.

Naphtha – a fraction of crude oil (see above) and the crucial component of plastic.

National Sword – an initiative from China to reduce the import of waste.

Near infrared (NIR) sorting – technology that detects differences in the wavelengths of infrared light that is reflected by polymers with different chemical structures to aid in sorting them.

Non-target material – material that is not wanted by the sorter or recycler.

OECD (Organisation for Economic Co-operation and Development) – a group of 38 countries, including the UK, brought together to stimulate economic development and world trade.

Ore – a naturally occurring rock from which metal can be extracted.

Overhead magnet – a magnet placed over a conveyor belt, used to extract magnetic material (like steel), separating it from other waste.

Oxide – a compound of oxygen with another element.

Oxo-degradable – a plastic type with an additive which causes the plastic to break down into fragments.

Packaging – any material used to hold, protect, handle, deliver or present goods.

Paperboard – a thick paper-based material, for example the cardboard used to make a cereal box.

Paper mill – a factory in which paper is manufactured.

Paris Agreement – a legally binding agreement on climate change.

Pay as you throw – the concept of charging residents for the amount of material they throw away, usually waste destined for landfill.

Pellet – a small sphere, usually as the output of a plastic recycling process.

PET – a plastic type most commonly used for drink bottles.

Pig iron – the result of smelting iron ore in a furnace, before it is turned into steel.

Plastic Pact – an agreement formed in multiple countries with many companies who commit to meet ambitious sustainability and recycling targets.

Polymer – a substance made up of multiple monomers (see above); plastic is a common polymer.

Polymerisation – the process by which many monomers come together to form a polymer.

Polyolefin – a polymer produced from a simple olefin as its monomer; polyethylene and polypropylene are both polyolefins.

Portable battery – a battery weighing less than 4 kg, which is not automotive or industrial.

PP – polypropylene, a plastic made from many propylene monomers and used in a lot of food contact applications.

PRF (Plastics Recycling Facility) – like a MRF (see above), designed specifically to separate different plastic types.

Primary packaging – the packaging around an individual product, for example, a wrapper around a bar of chocolate.

Producer – a company responsible for placing products on the market: this could be a manufacturer, brand, retailer or importer.

Producer responsibility – legislation which requires companies to pay for the recycling of their products at the end of their life.

PS (polystyrene) – a plastic type, found in expanded or rigid form, used in many packaging applications.

Pulp – a material which results from chemical or mechanical separation of cellulose fibres from wood.

Purification recycling – a form of chemical recycling, dissolving plastic in a solvent and then, through a purification process separating it from any non-target material, creating a pure polymer.

PVC – a weather-resistant plastic type, used in pipes and the plastic often used to make cling film.

Pyrolysis – the process of breaking down plastic polymers with heat.

Raw material – the basic material from which a product is made.

Recyclate – raw material which has been sent to a recycling plant to be processed.

Recycled content – the part of packaging or a product made from recycled material.

Recycling – the action or process of converting waste into a reusable material.

Recycling rate – the amount of recycling taking place for a given material.

Refinery – a factory where substances are purified.

Regrind – plastic which has been mechanically shredded to a smaller size.

Rigid plastic – plastic that is hard, used for things like bottles, tubs and trays.

Scrunch test – a test to check whether a product is made primarily from paper or is actually plastic coated in metal: simply scrunch the material into a ball in your hand, then open your hand – if it has kept its shape, it can be recycled.

Secondary packaging – the packaging around multiple units, like a set of rings holding multiple drink cans.

Shredder – a device to shred plastic as it passes through the machine.

Single-use plastic – plastic which is designed for only one use before being disposed of.

Sink-float tank – a tank used to separate different materials based on their specific weights.

Specialist collection – a collection not taking place at the kerbside or civic amenity site, for example a community collection point.

Supply chain – the sequence of companies or processes involved in the production of a product.

Sustainable Development Goals – seventeen interlinked global goals designed to create a more sustainable future for all.

Tertiary packaging – the packaging used to transport secondary and primary packaging (see above), usually a pallet or outer box.

Thermal runaway event – an incident where one thermal process causes another.

Tidyman – an icon found on product packaging which encourages people to dispose of the packaging after use.

Two-stream collection – a collection involving two bins, one for mixed plastic, metal, cartons and glass and the other focused on paper and card.

Waste – any substance or object which the holder discards, intends to discard or is required to discard.

Waste carrier – a business registered to collect and move waste.

Waste hierarchy – the order waste should be managed, from prevention (most preferred) to landfill (least preferred).

Waste stream – flows of specific waste from disposal to recycling or recovery.

Waste transfer note – a document that details the transfer of waste.

Wishcycling – also known as 'aspirational recycling' (see above), a consumer putting something in the recycling bin that cannot be recycled.

Wooden pallet – a flat wooden structure to transport products.

Notes

Chapter 1: The Rubbish Basics

1   https://www.lexico.com/definition/recycling
2   https://www.legislation.gov.uk/eudr/2008/98/article/3
3   https://www.gov.uk/guidance/packaging-producer-responsibilities
4   https://www.livescience.com/28865-aluminum.html
5   https://alupro.org.uk/industry/local-authorities/environmental-benefits/
6   https://www.mazumamobile.com/why/about
7   https://www.un.org/sustainabledevelopment/sustainable-development-goals/
8   https://sdgs.un.org/goals
9   https://sdgs.un.org/goals/goal3
10 https://sdgs.un.org/goals/goal4
11 https://sdgs.un.org/goals/goal6
12 https://sdgs.un.org/goals/goal7
13 https://sdgs.un.org/goals/goal8
14 https://sdgs.un.org/goals/goal9
15 https://sdgs.un.org/goals/goal11
16 https://sdgs.un.org/goals/goal12
17 https://sdgs.un.org/goals/goal13
18 https://sdgs.un.org/goals/goal14
19 https://unfccc.int/process/the-paris-agreement/status-of-ratification
20 https://wrap.org.uk/taking-action/plastic-packaging/the-uk-plastics-pact
21 https://archive.wrap.org.uk/sites/files/wrap/The-UK-Plastics-Pact-report-18-19.pdf
22 https://wrap.org.uk/resources/report/eliminating-problem-plastics

23 https://blog.collinsdictionary.com/language-lovers/etymology-corner-collins-word-of-the-year-2018/

24 https://www.valpak.co.uk/more/material-flow-reports/drinks-recycling-on-the-go

25 https://committees.parliament.uk/work/1048/next-steps-for-deposit-return-schemes/news/139235/mps-examine-how-deposit-return-schemes-can-improve-plastics-recycling/

Chapter 2: The Rubbish History

1 https://www.recyclenow.com/recycling-knowledge/how-is-it-recycled/cans

2 https://alupro.org.uk/sustainability/fact-sheets/recycling-process-rate/

3 https://www.thetimes.co.uk/article/steels-last-stand-q928wpd9hvx

4 https://patentimages.storage.googleapis.com/ba/a8/14/d9d8dc4a5bbac3/US1508183.pdf

Chapter 3: The Rubbish Collection

1 https://www.royalmail.com/sites/default/files/royal-mail-prohibited-and-restricted-items-nov-23-2018---23410530_updated%20April%2019.pdf

2 https://www.valpak.co.uk/news-blog/blog/deposit-return-scheme-learning-from-the-norwegian-model

3 https://www.unep.org/news-and-stories/press-release/un-report-time-seize-opportunity-tackle-challenge-e-waste

4 https://npwd.environment-agency.gov.uk/FileDownload.ashx?FileId=94f324af-33a3-4473-83af-602a63ad8f88

5 https://npwd.environment-agency.gov.uk/FileDownload.ashx?FileId=d024140a-5b2d-40e6-9fb8-8133004562f5

Chapter 4: The Rubbish Sort

1 https://www.chemeurope.com/en/encyclopedia/Materials_recovery_facility.html

2 https://npwd.environment-agency.gov.uk/FileDownload.ashx?FileId=39e951c3-82fc-4b70-8397-74c6abdd7e0d

3 https://en.wikipedia.org/wiki/Alkaline_battery
4 https://archive.wrap.org.uk/sites/files/wrap/Choosing%20the%20
   right%20recycling%20collection%20system.pdf

Chapter 5: The Rubbish Recycling

1 https://www.recyclenow.com/recycling-knowledge/how-is-it-
   recycled/glass
2 https://npwd.environment-agency.gov.uk/
3 https://alupro.org.uk/industry/local-authorities/environmental-benefits/
4 https://www.recycle-more.co.uk/recycling/steel
5 https://www.recyclenow.com/recycling-knowledge/how-is-it-
   recycled/electricals
6 https://www.bbc.com/future/article/20161017-your-old-phone-is-
   full-of-precious-metals
7 https://www.statista.com/statistics/330695/number-of-
   smartphone-users-worldwide/
8 https://www.ecosurety.com/news/ecosurety-and-hubbub-launch-
   bringbackheavymetal-battery-collection-campaign/

Chapter 6: The Rubbish Knowledge

1 https://www.bbc.co.uk/news/science-environment-42953038
2 https://www.bbc.co.uk/news/business-43724314
3 https://wrap.org.uk/resources/guide/hdpe-plastic-bottles
4 https://www.dssmith.com/recycling/insights/blogs/2018/7/closing-
   the-loop-on-black-plastic-ready-meal-trays
5 https://publications.parliament.uk/pa/cm201617/cmselect/
   cmenvaud/179/17906.htm

Chapter 7: The Rubbish Symbols

1 https://www.pro-e.org/proe-members
2 https://www.saveonenergy.com/uk/recycling-symbols/
3 https://resource.co/article/keep-britain-tidy-re-launches-tidyman-
   symbol-11653
4 https://us.fsc.org/en-us/who-we-are/our-history

Chapter 8: The Rubbish Encyclopedia

1 https://www.cardfactoryinvestors.com/what-we-do/the-greeting-card-market

2 https://www.recycle-more.co.uk/household/recycling-facts

3 https://wrap.org.uk/resources/report/carrier-bag-use-and-attitudes

4 https://publications.parliament.uk/pa/cm201719/cmselect/cmenvaud/657/65705.htm

5 https://www.thegrocer.co.uk/hot-beverages-report-2016/nearly-a-third-of-brits-own-a-coffee-pod-machine/542542.article

6 https://www.sciencedirect.com/topics/engineering/cotton-farming

7 http://www.ace-uk.co.uk/media-centre/news/eu-beverage-carton-recycling-rate-hits-51/

8 https://zerowasteeurope.eu/2020/12/press-releasebeverage-carton-recycling-rates-substantially-lower-than-reported/

9 http://www.ace-uk.co.uk/media-centre/news/eu-beverage-carton-recycling-rate-hits-51/

10 https://wrap.org.uk/sites/default/files/2021-06/Food%20Surplus%20and%20Waste%20in%20the%20UK%20Key%20Facts%20June%202021.pdf

11 https://www.bedfed.org.uk/wp-content/uploads/NBF-National-Mattress-End-of-Life-Report-2019.pdf

12 https://www.bbc.co.uk/news/uk-45732371

13 https://www.thetimes.co.uk/article/billions-of-food-pouches-go-to-landfill-c7hf6lwc6

14 https://www.theguardian.com/commentisfree/2020/aug/07/coronavirus-britain-sandwich-pandemic-lunchtime

15 https://www.nationalgeographic.co.uk/environment-and-conservation/2018/02/straw-wars-fight-rid-oceans-discarded-plastic

16 https://www.intelligentliving.co/we-are-flushing-our-forests-down-the-toilet/

17 https://www.etrma.org/wp-content/uploads/2021/05/20210520_ETRMA_PRESS-RELEASE_ELT-2019.pdf

18 https://www.eunomia.co.uk/reports-tools/plastics-in-the-marine-environment/

19 https://www.theguardian.com/environment/2020/jul/14/car-tyres-are-major-source-of-ocean-microplastics-study

Chapter 9: The Rubbish Problems

1 https://ourworldindata.org/food-ghg-emissions

2 https://www.independent.co.uk/climate-change/news/the-timeline-rubbish-collection-2175131.html

3 https://wrap.org.uk/sites/default/files/2021-03/Key%20 Findings%20from%20the%20Recycling%20Tracker%202020.pdf

4 https://www.eunomia.co.uk/lithium-ion-battery-waste-fires-costing-the-uk-over-100m-a-year/

5 https://npwd.environment-agency.gov.uk/Public/ PublicSummaryData.aspx

6 https://www.greenpeace.org/international/press-release/47759/ investigation-finds-plastic-from-the-uk-and-germany-illegally-dumped-in-turkey/

7 https://www.letsrecycle.com/news/latest-news/retailers-reduction-plastic-bags/

8 https://www.greenpeace.org.uk/wp-content/uploads/2021/01/ Checking-Out-on-Plastics-III-FINAL.pdf

9 https://assets.ctfassets.net/bffxiku554r1/4TNPmg4tzrgo H39Ievy7lC/763dce615ef1b49a50d592c38c199bb8/Coop-Bag-to-Rights-Report.pdf

10 https://assets.publishing.service.gov.uk/government/uploads/system/ uploads/attachment_data/file/291023/scho0711buan-e-e.pdf

11 https://www.eunomia.co.uk/reports-tools/plastics-in-the-marine-environment/

12 https://www.sciencedirect.com/science/article/abs/pii/ S0025326X16307639?via%3Dihub

13 https://www.eunomia.co.uk/reports-tools/plastics-in-the-marine-environment/

14 https://www.scientificamerican.com/article/stemming-the-plastic-tide-10-rivers-contribute-most-of-the-plastic-in-the-oceans/

15 https://theoceancleanup.com/sources/

16 https://journals.plos.org/plosone/article?id=10.1371/journal.
pone.0111913

17 https://www.eunomia.co.uk/reports-tools/plastics-in-the-marine-
environment/

18 https://lifewithoutplastic.com/plastic-in-oceans/

19 https://npwd.environment-agency.gov.uk/FileDownload.
ashx?FileId=df73ed69-249c-4fa3-906c-3b4ab197cf0a

Chapter 10: The ~~Rubbish~~ Better Future

1 http://www.news.pitt.edu/sites/default/files/documents/
TaboneLandis_etal.pdf

2 https://www.letsrecycle.com/news/latest-news/move-four-weekly-
collections-conwy/

3 https://www.forestresearch.gov.uk/tools-and-resources/statistics/
statistics-by-topic/timber-statistics/uk-wood-production-and-trade-
provisional-figures/

Index

Unbound is the world's first crowdfunding publisher, established in 2011.

We believe that wonderful things can happen when you clear a path for people who share a passion. That's why we've built a platform that brings together readers and authors to crowdfund books they believe in – and give fresh ideas that don't fit the traditional mould the chance they deserve.

This book is in your hands because readers made it possible. Everyone who pledged their support is listed below. Join them by visiting unbound.com and supporting a book today.

Patrons

Cromwell Polythene Ltd   Impact Solutions
Impact Recycling

Supporters

Andrew Adams   Emily Arch
Anna Aitken   James Armitage
Colin Alden   Steph Atkinson
Sally Allen   Emmeline Aves
Rhona Allin   James Aylett
Chris Alveyn   Thomas Baker
Mary Alveyn   Vicki Baker
Kaleigh Anstee   Karen Balmforth

Gianluca Balzamo

Gareth Barrett

Finn Bartram

Margaret Bates

Dave Beal

Rosemary Beardow

Roger Beck

Anna Beel

Claire Bees

Ronnie Bendall

Katy Bevan

Tessa Bircham

Andrew Bird

Beau Birkett

Peter Blackburn

Noreen Blanluet

Oana Bogdan

Simon Bostock

Dan Boyle

Isla Bradley

Julia Bragg

Alison Bramfitt

Bruce Bratley

Lucinda Brook

Alexander Brookes

Elizabeth Buckley

Heidi Budino

Erika Bunao

Claire Burden

Bev Burnham

David Burrows

Sammy Burt

Jo Caffrey

Lisa Cain

Emma Capello

Kathryn Chabarek

James Champ

Martin Chapman

Danny Cheke

Stephen Clark

Charles Clarke

Jayne Clementson

Rebecca Colley-Jones

Gemma Cowin

Christian Crawford

John Crawford

Team Crews Knowsley

Rory Cronin

Stamati Crook

John Crowther

Kathrine Cuccuru

E R Andrew Davis

Matthew Demmon

Ben Doran

Lucy Drake-Lee

Paul Driver

Louise Dudley

Jane Duncan

Steven Duncan

Martin Eggleston

Matthew Elford

Kat Emmett

Lisa Evans

Megan Evans

Patrick Fairclough

Alison Falzon

Helen Farmer

Justin Farrimond

Virginia Fassnidge

Devanie Fergus

Laura Fernandez

Ali Fisher

Alice Flavin

Sally Foote

Jean Forbes

Clare Fowler

Bowie Frances Jean Wilson

Susan Fraser

Lizzie and Nathan Fulton

Pippa Fulton

Sally Gale

Julie Gallacher

Clement Gaubert

Ray Georgeson

Mark Gethings

Alice L Gibbs

Simon F Gibbs

Daniele Gibney

Sukie Gladstone

Steffi Goetzel

Louisa Goodfellow

Heide Goody

Andrew Grant

Colm Grimes

Alina Gromova-Jones

Katy Guest

Ryan Hacker

George Hadfield

Alexandra Hanna

Brian Hanson

Samantha Hardwell

Matt Haskell

Katie Hatton

Cathy Henderson

Simon Hetzel

Amanda Hickling

Connor Hill

Catherine Hills

Steven Hinchly

Craig Hiscock

James HiScock

Mark Horley

Paul Horton

| brandprintcolour™

Steph Housty

Jennifer Hurstfield

Luke Hutchison

Martin Hyde

Sam Hyde-hart

Grace Jansen

Paul Jenkins

Roger Jones

Lauma Kazusa

Katrina Kelly

Ella Kennedy

Becca Ketley

Dan Kieran

Ellie King

Omer Kutluoglu

Pierre L'Allier

Nigel Lax

Catherine Leach

Melanie Leach

David Lee

Rod Leefe

Emilia Leese

Stuart Lendrum

Ryan Longstaff

Ben Luger

Rob MacAndrew

Catherine MacDonald

Peter Maddox

Matt Manning

Alexander Marrs

Deborah Marshall

James Marshall

Angela Martin

Hannah Martin

Leonnie Matthew-Morgan

Niki McCann

Julie McCarthy

Ewan McClymont

Alessandra McConville

Amber McGuigan

Robert McIntosh

Andrew McKenzie

Nicole McNab

Skylark Media

Rory Miles

Nicollyn Mitchell

John Mitchinson

Fergus Mooney

Jo Mooney

Andy Moore

Mark Morfett

Liz Morrish

Simon Morrison

Lucy Mortimer

Gareth Morton

Katrina Moseley

Lewis Mottashed

Gill Mulroe

Siobhan Murphy

David Newman

Jai Newton

Roseline Nicholls

Gary Nicol

Steve Noss

Carlo Novato

Anna O'Dell

Jesper Oelert-Pedersen

John Parker

Oliver Parker

Sanjay Patel

Ellen Pennifold

Dan Peters

Anthony Peyton

Andy & Georgia Piper

Ellie Piper

Louise Piper

Lucas Piper

Paul Pivcevic

Chris Platt

Justin Pollard

Neil Pollard

Luca Pornaro

Colin Porter

James Potten

Helen Potter

Carl Pratt

William Preston

Sean Price

Phil Prior

Catherine Pulman

Julie Pye

Doug "Dhomal" Raas

Ralph the Dog

JP Rangaswami

Colette Reap

Simon Reap

Tilly Redding

Sam Reeve

Paul Rendle-Barnes

Oliver Reynolds

Chris Richards

Sophie Rivett-Carnac

Däna Roberts

Keiron Roberts

Tony Roberts

Anthea Robertson

Lucy Robinson

Sarah Rogerson

Laure Rouaux

Lin Roussel

George Rumble

Charlotte Russell-Parsons

Deborah Sacks

Dominique Sandy

Benjamin Saunders

Pawan Saunya

Jason Savage

Matt Sawyer

Joe Scaife

Lesley Scarles

Laura Scott

Nigel Scott

Richard Selby

Dan Sewell

Steve Sheath

Ronnie Sievewright

Naveen Singh

Ian Smith

Vickesh Solanki

Richard Soundy

Adrian Spottiswoode

Wendy Staden

Robbie Staniforth

Tim Stevens

Jeremy Stewart

Susanne Stohr

Frankie Stone

Oliver Strudwick

Kris Sullivan

David Syrett

Christian Tait

Mark Talpade

Dionne Taylor

Pam Taylor

Bernie Thomas

Andy Thompson

Simon Tilling

Dylan Topham

Jon Treacher

Suzanne Trew

Martin Trotter

Andres Tunon

Scott Turner

Akane Vallery Uchida

Matthew Unerman

Alan Valentine

Zach Van Stanley

Vicky Vella

Antonia Venning

Kevin Vyse

Robert Wade

Tara Walker

Gareth Wallwork

Abigail Warren

Edward Warren

Liz Weldrake

Jo West

Henry Whitworth

Edward Whyte

Grace Whyte

Karen Whyte

Sarahjane Widdowson

Jamie Winter

Emma Wise

Greg Wood

Tom Wood

Stacey Woods
Josh Wytchard
Laura Yates
Julie Zeraschi